Pathways to a Sustainable Economy

Moazzem Hossain • Robert Hales
Tapan Sarker
Editors

Pathways to a Sustainable Economy

Bridging the Gap between Paris Climate Change Commitments and Net Zero Emissions

Editors
Moazzem Hossain
Griffith Asia Institute and Department of
International Business and Asian Studies
Griffith University
Nathan, Queensland, Australia

Robert Hales
Griffith Centre for Sustainable Enterprise
Griffith Business School
Griffith University
Nathan, Queensland, Australia

Tapan Sarker
Griffith Centre for Sustainable Enterprise
and Department of International Business
and Asian Studies
Griffith University
Nathan, Queensland, Australia

ISBN 978-3-319-67701-9   ISBN 978-3-319-67702-6 (eBook)
DOI 10.1007/978-3-319-67702-6

Library of Congress Control Number: 2017955005

© Springer International Publishing AG 2018
This work is subject to copyright. All rights are reserved by the Publisher, whether the whole or part of the material is concerned, specifically the rights of translation, reprinting, reuse of illustrations, recitation, broadcasting, reproduction on microfilms or in any other physical way, and transmission or information storage and retrieval, electronic adaptation, computer software, or by similar or dissimilar methodology now known or hereafter developed.
The use of general descriptive names, registered names, trademarks, service marks, etc. in this publication does not imply, even in the absence of a specific statement, that such names are exempt from the relevant protective laws and regulations and therefore free for general use.
The publisher, the authors and the editors are safe to assume that the advice and information in this book are believed to be true and accurate at the date of publication. Neither the publisher nor the authors or the editors give a warranty, express or implied, with respect to the material contained herein or for any errors or omissions that may have been made. The publisher remains neutral with regard to jurisdictional claims in published maps and institutional affiliations.

Printed on acid-free paper

This Springer imprint is published by Springer Nature
The registered company is Springer International Publishing AG
The registered company address is: Gewerbestrasse 11, 6330 Cham, Switzerland

Preface

The motivation for compiling this volume has stemmed from an historic global collective realisation that a new way of living is needed. Although many authors have written about climate change and sustainability, there is a significant difference between previous ways of understanding the linkage compared to the present. The new global agreement on climate change has coincided with the dawn of the new geological era—the Anthropocene. Feedback from nature, as a consequence of fossil fuel driven economic growth, is now being differentially experienced by people as a range of negative impacts on social and environmental systems. The recent global agreement on climate change signifies a shift in the way sustainability is conceived and practised. The imperative of sustainability has shifted from a "nice to have" concept to a "must have" practice if socioeconomic development is to be constrained within environmental limits.

The purpose of this volume, as articulated in the title and under the aims and objectives presented in the Introduction, is indeed unique with respect to making a case of promoting sustainable use of resources and bringing down pollution targets keeping the COP21 agenda in mind as agreed upon in Paris in December 2015. The primary audience of the volume would be the UN development agencies, academic institutions, government policymakers, NGOs and business leaders. It will focus on early adopters of change and technology with a view to achieve a sustainable economy in the twenty-first century. The book is based on the COP21 summit held in Paris in December 2015 and investigates the pressing issues regarding the gap between COP21 commitments of the UN member states and 2030 targets stipulated as the key measures for emission control. Offering critical perspectives on how to achieve the COP21 targets at the early stage of the agreement will facilitate the likely success in achieving these targets. Importantly, the investigation will be based on case studies from Australia and other parts of the Asia-Pacific to offer grounded perspectives of the critique.

Nathan, QLD, Australia	Moazzem Hossain
Nathan, QLD, Australia	Robert Hales
Nathan, QLD, Australia	Tapan Sarker

Acknowledgements

We would like to take this opportunity to thank the Griffith Centre for Sustainable Enterprise (GCSE) located under the Griffith Business School at Griffith University, Australia, for its generous support for organising the international conference "Pathways to a Sustainable Economy: Bridging the Gap between Paris Agreement Commitments and 2030 Targets in Brisbane in November 2016". Selected papers out of this conference have been included in this volume. Vanessa Taveras Dalmau of the GCSE has been providing all-out support for successful completion of the volume. Despite the challenges of communicating with the contributors based in different parts of Australasia, she put all-out efforts to meet the deadlines in preparing the manuscript. We also acknowledge the time and skill of our in-house editor Malini Devadas in making a multidisciplinary manuscript presentable to the readers.

We must appreciate the time and efforts of all the contributors for committing their time to the project and for sharing their research outcome and knowledge in this volume.

The Editors
Brisbane, July 2017

Contents

1 Introduction: Pathways to a Sustainable Economy 1
   Moazzem Hossain

Part I   Critical Perspectives on Achieving a Zero Emissions Future

2 Joining the Dots: Sustainability, Climate Change
   and Ecological Modernisation 15
   Michael Howes

3 A Systems Critique of the 2015 Paris Agreement on Climate............ 25
   Luke Kemp

4 Structural Impediments to Sustainable Development
   in Australia and Its Asia-Pacific Region 43
   Ahmed Badreldin

Part II Strategies to Achieve Sustainable Economy
          and Zero Emissions Future

5 The Role of Planning Laws and Development Control
   Systems in Reducing Greenhouse Gas Emissions: Analysis
   from New South Wales, Australia 61
   Nari Sahukar

6 Carbon Disclosure Strategies in the Global Logistics
   Industry: Similarities and Differences in Carbon Measurement
   and Reporting 87
   David M. Herold and Ki-Hoon Lee

7 Synergy between Population Policy, Climate Adaptation
   and Mitigation 103
   Jane N. O'Sullivan

Part III Major Challenges Towards a Sustainable Future: Case Studies from Asia

8   From "Harmony" to a "Dream": China's Evolving Position on Climate Change 129
    Paul Howard

9   COP21 and India's Intended Nationally Determined Contribution Mitigation Strategy 149
    Ranajit Chakrabarty and Somarata Chakraborty

10 Seasonal Drought Thresholds and Internal Migration for Adaptation: Lessons from Northern Bangladesh 167
    Mohammad Ehsanul Kabir, Peter Davey, Silvia Serrao-Neumann, and Moazzem Hossain

11 Incentives and Disincentives for Reducing Emissions under REDD+ in Indonesia 191
    Fitri Nurfatriani, Mimi Salminah, Tim Cadman, and Tapan Sarker

12 Carbon Budgeting Post-COP21: The Need for an Equitable Strategy for Meeting CO_2e Targets 209
    Robert Hales and Brendan Mackey

Index 221

About the Authors

Ahmed Badreldin is studying postgraduate research at the University of Newcastle in Australia. In January 2016, his article "Energy Crisis Keeps Egypt on the Wrong Side of Capitalism" earned him both the International Award for Excellence for Volume 8 of *the Global Studies Journal* and Graduate Scholar Award from Common Ground Publishing. He has a high school diploma from Gainesville High School in Georgia, USA; a B.Acc. from Ain Shams University in Egypt; an MBA from Edinburgh Business School, Heriot-Watt University in Great Britain; as well as almost 7 years of banking experience. Ahmed lives in Newcastle, Australia.

Tim Cadman, B.A., M.A., Ph.D. is a Research Fellow in the Institute for Ethics, Governance and Law at Griffith University, Queensland, Australia. He is also a Senior Research Fellow in the international Earth Systems Governance Project and an Adjunct Research Fellow and Member of the Australian Centre for Sustainable Business and Development at the University of Southern Queensland. He specialises in the governance of sustainable development, climate change, natural resource management including forestry, and responsible investment.

Ranajit Chakrabarty, M.Sc., Ph.D. (Statistics) was the head of the Department of Business Management, Calcutta University, where he served for 40 years. He was also the Dean, Faculty of Commerce, Social Welfare and Business Management, Calcutta University, in 2002–2004 and 2005–2007. His research interest covers areas ranging from operational research, quantitative methods, econometrics, environment management and marketing. Dr. Chakrabarty received the Russell Ackoff Award at the 19th International Conference on Solid Waste Technology and Management, Philadelphia, USA, in 2004 and received the "Excellent Paper" award at the 9th China International Conference, Beijing University, China, in 2009.

Somarata Chakraborty, P.G.D.M. (Finance), M.Sc. (Eco) is currently pursuing PhD from Calcutta University and has 9 years of teaching experience and 3 years of experience in Industry. The area of research includes industrial organisation and environment management, and Somarata is working as Assistant Professor in Camellia Institute of Technology under MAKAUT. Prior to joining the Camellia group, Somarata served at the Narula Institute of Technology, JIS Group, for 4 years also associated with Army Institute of Management, Kolkata, Future Innoversity (Future Learning Initiative), Camellia School of Business Management, etc. as Visiting Faculty. Somarata has published 4 articles in national and international journals and presented papers at several seminars and conferences.

Peter Davey is Program Director of the Bachelor of Environmental Management and specialises within the Master of Environment in environmental protection at Griffith University, Brisbane. Peter teaches and researches environmental health topics internationally and focuses on the planning for and management of climate change impacts and disaster risk reduction and implications for sustainable livelihoods. Peter is an Accredited UNISDR Disaster Trainer and conducts intensive short courses for developing country professionals and industry.

Robert Hales is the Director of the Griffith Centre for Sustainable Enterprise at the Griffith Business School. Rob has led on climate change research projects including developing the Business Case for Climate Change Adaptation for the National Climate Change Research Facility. He has also recently partnered with the Global Change Institute to deliver policy advice to the Climate Action Round Table. Rob plays a key role in the Business School, facilitating impactful research on the topic of sustainability. He currently teaches in the MBA programme delivering the course Sustainability and Systems Thinking.

David M. Herold is a Sustainability Researcher with a focus on Decarbonising Transport and Entrepreneurial Innovation. He is currently pursuing his PhD which examines the implementation of sustainability initiatives and their implications within the logistics and transportation industry. He holds a BA in Business Administration, an MBA degree and an MA in International Relations and has held visiting and teaching positions across Asia-Pacific and Europe. Prior to his academic career, David worked for more than 10 years in a Fortune 500 Global Logistics company and has extensive industry and management experience in strategic planning and business development.

Ki-Hoon Lee is an internationally recognised leader in the field of corporate sustainability management. Through his research, Professor Lee embraces environmental, social and economic challenges, integrating sustainability into the business value chain to enhance both business and societal value. He has a strong research focus on corporate sustainability management, in particular strategic management and corporate sustainability, carbon management and business strategy, green and sustainable supply chain management, and sustainability management accounting. His recent book (with Stephan Vachon) *Business Value and Sustainability* is published in 2016 (with Palgrave, London).

Moazzem Hossain is an Adjunct Associate Professor of the Department of International Business and Asian Studies at Griffith University, Australia. Over the last three decades, his research has covered forestry economics, economic development, telecommunications regulation, and climate change and growth in Asia. Recently, he co-edited the volume *South-South Migration: Emerging Patterns, Opportunities and Risks* published by Routledge, London and New York, in April 2017.

Paul Howard is currently the Program Director for the Bachelor of Asian Studies and a Lecturer within the Department of International Business and Asian Studies, Griffith Business School. Paul is also a member of the Griffith Asia Institute. Along with the focus on development and policy in China, Paul's research interest extends to the political economy of other developmental issues such as sanitation development in India and Cambodia. More broadly, his particular research interest is on the interplay between formal policy and actual outcomes for citizens in developing economies.

Michael Howes is the Deputy Head of the Griffith School of Environment and a researcher with the Cities Research Institute at Griffith University, Australia. His work explores how governments try to make society more sustainable and resilient, with specific projects on climate change, sustainable development, environment protection, public environmental reporting and ecological modernisation. Before becoming an academic, Michael worked for several years as an industrial chemist and technical manager in the manufacturing sector. He has also been a member of the Queensland Conservation Council board and chaired a Technical Advisory Panel for the Australian Government's National Pollutant Inventory.

Mohammad Ehsanul Kabir is currently a PhD candidate at the Griffith School of Environment, Griffith University, Australia. Ehsan's PhD project involves unpacking vulnerabilities of internal migrants living in the northwestern drought-prone area of Bangladesh. Ehsan earned his master's degree from Monash University, Australia, in international development and environmental analysis and a bachelor's degree in social science from Independent University Bangladesh. Ehsan has published journal articles on political ecology, disaster risk reduction, environmental conflict and internal displacement. He is also an Assistant Professor at the Dhaka School of Economics, a constituent institution of the University of Dhaka.

Luke Kemp is a lecturer in climate and environmental policy at both the Fenner School of Environment and Society and Crawford School of Public Policy at the Australian National University (ANU), and Senior Economist with Vivid Economics. He holds both a Doctorate in Political Science (2016) and a Bachelor of Interdisciplinary Studies with first class honours from the ANU (2011). His research looks at why international agreements succeed and fail, with a particular focus on the role of the United States. He is a regular media commentator and his research has been covered by international media outlets such as the *Washington Post* and *New York Times*.

Brendan Mackey is Director of the Climate Change Response Program at Griffith University, Australia. He has a PhD from The Australian National University in ecology. Brendan has authored and co-authored over 200 academic publications. His current research is focussed on ecosystem-based approaches to mitigation and adaptation in the context of sustainable development.

Fitri Nurfatriani is a researcher at the Research and Development Center for Social Economics, Policy and Climate Change in the Ministry of Environment and Forestry of the Government of Indonesia. She is actively involved in various research activities providing science-based policy recommendations for the ministry. Her research areas are economic valuation of forest resources, forest economics and policy, forest fiscal policy and institutional climate change. Fitri holds a bachelor's degree in forestry, and a master's and PhD degree in forest science from Bogor Agricultural University Indonesia.

Jane N. O'Sullivan is a former senior researcher at the University of Queensland's School of Agriculture and Food Sciences, where she led research programmes on agricultural intensification of subsistence crops in the Pacific and Southeast Asia. She subsequently turned attention to the demographic pressures on food security and economic development. She has published groundbreaking work on the economic impacts of population growth, the "demographic dividend" and population ageing.

Nari Sahukar The Environmental Defenders Office New South Wales (EDO NSW) is a non-government community legal centre that has helped the community protect the environment through law since 1985. Nari Sahukar is a Senior Policy and Law Reform Solicitor at EDO NSW, with over 10 years' public policy and legal expertise in and outside government. He assists community and environment groups, government agencies and lawmakers to improve environmental regulation at local, state and national levels. His current practice areas include planning and development law, mining law, pollution, biodiversity protection and climate change and energy law. Nari has a Bachelor of Arts and Bachelor of Laws (Hons 1) from Macquarie University and a Master of Political Economy from the University of Sydney.

Mimi Salminah is a researcher at the Research and Development Center for Social Economics, Policy and Climate Change under the Ministry of Environment and Forestry, Government of Indonesia. She is actively involved in various research activities providing scientific-based policy recommendations for the ministry. Her research areas are forest landscape management, forest hydrology, forest policy and climate change. Mimi holds a bachelor's degree in forestry from Bogor Agricultural University, Indonesia, and a master's degree in forest science and management from Sothern Cross University Australia.

Tapan Sarker is a Senior Lecturer in the Department of International Business and Asian Studies and a member of the Griffith Asia Institute and Griffith Centre for Sustainable Enterprise. He is a former World Bank scholar. His most recent co-authored book publications are *Political Economy of Sustainable Development* (Edward Elgar, 2015) and *The Asian Century, Sustainable Growth and Climate Change: Responsible Futures Matter* (Edward Elgar, 2013).

Silvia Serrao-Neumann is a Senior Research Fellow for the Corporate Research Centre for Water Sensitive Cities at the Cities Research Institute, Griffith University, Australia. Her research focuses on climate change adaptation from multiple perspectives, including catchment scale landscape planning for water sensitive city regions; cross-border planning and collaboration; disaster recovery under a stakeholder-focused collaborative planning approach; natural resource management; and action/intervention research applied to planning for climate change adaptation.

Abbreviations

BADC	Bangladesh Agricultural Development Corporation
BMDA	Barind Multi-purpose Development Authority
CCP	Chinese Communist Party
CDP	Carbon disclosure project
COP	Conference of Parties
CPO	Crude palm oil
DR	Dana Reboisasi
EIA	Environmental impact assessment
EIS	Environmental impact statement
EM	Ecological modernisation
EPA	Environment Protection Authority
EPL	Environment protection licence
FFI	Fossil fuel and industry
GCF	Green Climate Fund
GDP	Gross domestic product
GRT	Ganti Rugi Tegakan
IIUPH	Iuran Izin Usaha Pemanfaatan Hutan
INDC	Intended nationally determined contribution
IPCC	Intergovernmental Panel on Climate Change
IPPKH	Ijin Pinjam Pakai Kawasan Hutan
ISDS	Investor-state dispute settlement
LEP	Local environmental plans
LULUCF	Land use change and forestry
MDG	Millennium Development Goals
MNRE	Ministry of New and Renewable Energy
MoEF	Ministry of Environment and Forestry
NDC	Nationally Determined Contributions
NDRC	National Development and Reform Commission
NGO	Nongovernment organisation
NSW	New South Wales
NTSR	Non-tax state revenue

PBB	Pajak Bumi Bangunan
PET	Population-energy-technology
PRC	People's Republic of China
PSDH	Provisi Sumber Daya Hutan
RED	Reducing emissions from deforestation
REDD	Reducing emissions from deforestation and forest degradation
SEPP	State environmental planning policies
SSP	Shared socioeconomic pathways
TFR	Total fertility rate
UN	United Nations
UNFCCC	United Nations Framework Convention on Climate Change
US	United States

Chapter 1
Introduction: Pathways to a Sustainable Economy

Moazzem Hossain

Abstract The purpose of this volume, as articulated in the title and under the aims and objectives presented in this introductory chapter, is indeed unique with respect to making a case of promoting sustainable use of resources and bring down pollution targets keeping the COP21 agenda in Paris in mind. It is now established that climate change has affected every continent, from the equator to the poles, from the mountains to the coasts. The many kinds of extremes that climate change has contributed so far are heat waves, heavy rain and untimely floods, wildfires, drought and melting ice and snow contributing to sea level rise. This volume is based on a conference held in Brisbane, Australia in November 2016—1 year on from COP21. The main objective of this conference was to identify the key issues that need to be addressed in order for nations to achieve their Intended Nationally Determined Contributions (INDCs) proposed at COP21. With this in mind, this volume has focused on how one can add further to the present debate on emission control globally involving both developed and developing nations of the Asia-Pacific. This introductory chapter gets the debate to a level where it will be clear to readers how the aims and objectives of the volume can be achieved.

The focus of this volume, thus, is on identifying solutions to the challenges facing governments, businesses and civil society from countries in the Asia-Pacific region in meeting mitigation targets. In other words, the aims are to: offer critical perspectives on the COP21 agreement and the challenges of meeting the targets; offer solutions to challenges of achieving a sustainable economy using climate change mitigation as a driver for change; and provide cases in Australia and Asia that can exemplify how such challenges can be faced. The overall outcome of the volume suggests that there are considerable challenges for nations to achieve the goals of the Paris Agreement in 2015. In this volume, the efforts of the contributors were to analyse how to bridge the gap between COP21 commitments and targets of emissions control while employing an interdisciplinary approach.

Keywords UNFCCC • Climate change • UNESCO HQ • COP21 • Net zero emissions future • Sustainable economy • Asia-Pacific

M. Hossain (✉)
GAI and IBAS, Griffith Business School, Griffith University, Nathan, QLD, Australia
e-mail: m.hossain@griffith.edu.au

© Springer International Publishing AG 2018
M. Hossain et al. (eds.), *Pathways to a Sustainable Economy*,
DOI 10.1007/978-3-319-67702-6_1

1.1 Background

The prosperity of the modern world over the second half of the twentieth century has been phenomenal for both advanced and emerging economies. In particular, populous developing economies like China and India to a great extent and South Africa and Brazil to a lesser extent have been achieving a rate of prosperity much faster than advanced nations, which had taken almost two centuries. According to many, however, advanced nations have reached full capacity; thus, the major aim of these economies is to maintain and sustain their prosperity in the current era of uncertainty in both global security and economic terms. Additionally, it cannot be denied that the present global uncertainty comes at a cost to the planet in terms of global warming-induced climate change. This is now a serious business for both developed and developing nations alike.

In view of the above, the United Nation (UN) through two major programmes has been attempting to address the problems of climate change and building sustainable economies: the UNFCCC's Conference of Parties (COP) initiatives and the recent declaration of the Sustainable Development Goals plan from 2016 to 2030 involving UN member nations. Both agendas have targets to be met by 2030. In this volume, we work closely with the UNFCCC agenda on climate change, with a view to identifying pathways to a sustainable economy in the medium to long term.

It is now established that climate change has affected every continent, from the equator to the poles, from the mountains to the coasts. The many kinds of extremes that climate change has contributed so far are heat waves, heavy rain and untimely floods, wildfires, drought and melting ice and snow. Leading to the COP21 in Paris in December 2015, the world's scientific community organised a 4-day conference under the umbrella of "Our Common Future under Climate Change" at the UNESCO HQ in July 2015. This conference was attended by nearly 2000 delegates from almost 100 countries to address the issues of adaptation, mitigation and sustainable development solutions due to global warming induced by the emission of CO_2 and other agents into the atmosphere from non-renewable anthropogenic energy sources. The first named editor of this volume was invited by the conference organisers to coordinate a session on the topic of "Land Adaptation in Asia".

This volume is based on a conference held in Brisbane, Australia in November 2016—1 year on from COP21 summit. The main objective of this conference was to identify the key issues that need to be addressed in order for nations to achieve their Intended Nationally Determined Contributions (INDCs) proposed at COP21. There are considerable challenges for each nation to achieve the overall goal of the Paris Agreement 2015. Members agreed that "If greenhouse gas emissions are cut by 40–70% below current levels by 2050, warming can be kept below 2 °C at reasonable cost by 2100".

With this in mind, this volume has focused on how one can add further to the present debate on emission control globally involving both developed and developing nations of the Asia-Pacific. This introductory chapter gets the debate to a level where it will be clear to readers how the aims and objectives of the volume can be achieved. Each chapter will include several recommendations for the future, which will be summarised and rationalised in the conclusion of the chapter. Each chapter will address one or more of the following issues: restoration, mitigation, adaptation and/or promoting resilience in the face of climate change as part of achieving a sustainable economy.

Since the volume is an edited edition, it is multidisciplinary in nature with chapters selected from various disciplines of science, engineering, business, environment and policy. The papers have been selected by the editors while keeping the objectives of the book in mind.

1.2 Premise

The main outcome of COP21 was the Paris Agreement 2015, which aims to strengthen the global response to the threat of climate change in the context of sustainable development and efforts to eradicate poverty. The Agreement commits the world's nations to take mitigation actions that will limit global warming to well below 2 °C above pre-industrial levels and pursue efforts to limit the temperature increase to 1.5 °C. Each nation's mitigation actions and target are specified in an INDC. A global stocktake will be undertaken at 5-year intervals to determine whether national mitigation actions in aggregate are on track to achieve the goal of limiting global warming to 1.5–2 °C. If not, nations have agreed to increase the ambition of their mitigation commitments.

Furthermore, nations have agreed to undertake rapid reductions to achieve a balance between anthropogenic emissions by sources and removals by sinks of greenhouse gases in the second half of this century, the period within which global greenhouse gas emissions must be reduced to zero if we are to limit warming to the 1.5–2 °C goal. For all countries, there is limited time to bridge the gap between their promises to take further action in mitigating greenhouse gas emissions, especially CO_2, and implementing the policies and programmes needed to ensure that the economy develops in ways that do not "cook the planet". The Paris Agreement stresses the importance of "non-state actors", especially the private sector, in achieving mitigation targets.

The focus of this volume, thus, is on identifying solutions to the challenges facing governments, businesses and civil society from countries in the Asia-Pacific region in meeting mitigation targets. In other words, the aims are to:

1. Offer critical perspectives on the COP21 agreement and the challenges of meeting the targets
2. Offer solutions to challenges of achieving a sustainable economy using climate change mitigation as a driver for change
3. Provide cases in Australia and Asia that can exemplify how such challenges can be faced

In order to address the aims, the volume is presented in three parts with this introductory statement on growth and sustainability. These are:

- Part I: Critical perspectives on a sustainable economy under a zero-emission goal
- Part II: Strategies to achieve a zero emissions future using case studies
- Part III: Major challenges towards a sustainable future: Case studies from Asia

It is universally accepted now that sustainability is the game changer for both developing and more developed nations in the twenty-first century. The Princeton University Press recently published a volume entitled Pursuing Sustainability, which presents thought-provoking issues on the subject of sustainability (Bruun & Casse, 2013; Matson et al. 2016). The book made it clear that it will help support teaching and research that "deals with sustainability in particular sectors such as energy, food, water, and cities, or in particular regions of the world". Among other things, this volume helps "in working collaboratively in governance processes to influence how society takes actions to promote sustainability". Two issues are important in this respect: unlocking sustainable good governance process of a nation; and society's actions towards promoting sustainability. Perhaps in these regards, the interdisciplinary nature of sustainable development could play a major part in the prosperity or wealth creation of nations.

Sustainable growth and prosperity (in other words, a sustainable economy) is a much talked-about area at present. Sustainable growth is important in the transition period of prosperity of a nation. What is transition? It means we left our backward past behind, are enjoying relative prosperity now and are looking forward to a prosperous future. All nations have gone through such a process since the industrial revolution more than 200 years ago. According to many development experts, China, India and other large nations, including former states of the Soviet Union, are in transition now (Gibson, 2016; Giddens, 2009).

All nations have been looking forward to attaining prosperity in this era of globalisation, information and digital revolution. One cannot deny the fact that sustainability has become the major issue for all transition and advanced states alike, including countries in the Asia-Pacific region and beyond. In terms of society's actions towards promoting sustainability and sustainable development, it is important to first define sustainable development.

There are many ways sustainable development can be defined. The most commonly used one is that from the Report for World Commission on Environment and Development in 1987: "Sustainable development is development that meets the needs of the present without compromising the ability of future generations to meet their own needs" (Ensor, 2011; Klein, 2015). This early definition contains two important concepts: first, the concept of needs, which underlines that the world should prioritise the needs of the poor; and second, the idea of limitations, which reminds the world that it should use resources available in the environment with technology and social organisation in a responsible manner. As we know, after all those years, the UN from January 2016 introduced a global programme called "Sustainable Development Goals" with 17 goals to 2030, endorsed by 200 nations.

In modern times, global higher education, development and training is the most important element of any society, and not only in giving or transferring knowledge. It is also core to the learning of all societies and its progress in applying scientific knowledge to new ventures, keeping in mind the costs and benefits to society and the nation at large (Bruun and Casse, 2013). In estimating both social and economic costs and benefits of new innovations and discovery, science, social science and other relevant disciplines play a paramount role (Berkes et al. 2003; Pelling, 2011). As mentioned earlier, the UNFCCC and other stakeholders have, since the early 1990s, put huge investment globally into addressing the issue of global warming-induced climate change for achieving sustainable and responsible growth of nations, developed and developing alike. Most importantly, the COP21 outcome in Paris in 2015 gives major hope towards achieving this goal. In other words, the time has come to work for a sustainable economy goal across transition and advanced nations. Since this volume aims to find pathways of a sustainable economy, one must make this term clear upfront: what does a sustainable economy mean?

Having said this, there are many obstacles remaining in order to work effectively towards the goal of attaining a sustainable economy in the medium to long run. In view of the above, the individual studies in this volume make some effort to investigate pathways to a sustainable economy in view of the outcome of the COP21.

1.3 Summary of Chapters

1.3.1 Part I: Critical Perspectives on Achieving a Zero Emissions Future

At the 1992 Rio Earth Summit, the governments of the world agreed, among other things, to address climate change and pursue sustainable development. After more than two decades of disappointing progress, two new sets of objectives were created in 2015 by the Paris Agreement on climate change, and in the United Nations Sustainable Development Goals. It is hoped that these will reinvigorate action on both issues, but the challenge is how to turn such aspirations into the tangible on-the-ground changes needed. The challenge is made even more difficult by strong opposition from some sectors, competing priorities and a highly charged political arena. Chapter 2 explores the option of using the idea of ecological modernisation to develop an integrated, multipronged strategy that could win the support of governments, business and the community. Such a strategy would involve promoting technological innovation, engaging with economic imperatives, political and institutional change, transforming the role of social movements and a discursive change in the way both the problems and solutions are perceived. Nowhere is such a strategy needed more than in the rapidly industrialising Asia-Pacific region, which has become a manufacturing hub for the world. While industrial development has brought some benefits, it has also come at the cost of rising greenhouse gas emissions, increasing pollution, growing resource depletion and the loss of biodiversity.

Decoupling economic growth from such environmental damage in this region should therefore be a priority. This chapter argues that while the idea of ecological modernisation arose in Europe, it can be adapted to the Asia-Pacific region in order to create a strategy that generates the necessary environmental, economic and social benefits.

Chapter 3 provides a system dynamics critique of the Paris Agreement. It puts forward a "multilateral systems" theory and framework to analyse international agreements as complex systems that drive change over time. This framework is then used to critically explore why the Paris Agreement will or will not work over time. The gap between existing targets and trajectories required for meeting global goals can only be met through significant increases in ambition. The logic of the Agreement is to heighten ambition over time through increasing public and peer pressure, and delivering a "signal" to markets that triggers low-carbon investments. Both of these reasons are, in the language of systems thinking, examples of "positive feedback" that the Paris Agreement could trigger. However, neither the pressure nor signal arguments have a strong basis in either empirical evidence or historical precedent. The Paris Agreement relies on unproven mechanisms to increase change over time: it is a multilateral experiment. Moreover, there are significant balancing feedbacks, such as carbon lock-in dynamics, that will likely stifle action over time. Unless other stronger feedbacks are activated, such as through the use of carbon clubs and trade measures, then the Paris Agreement is unlikely to achieve its goals.

The primary objective of Chap. 4 is to investigate structural impediments that confront and prevent Australia and the Asia-Pacific region from achieving their Paris Agreement targets and consolidating sustainable development. While neoliberal globalisation has nurtured ecological damage and widespread poverty and wealth inequalities in a systematic manner, this chapter argues that the accumulation and persistence of these structural deficiencies portend severe implications against the attainment of sustainability targets. The chapter introduces an assessment approach, suggesting the stage of economic development, social equity and political orientation of each country distinguishes its vulnerability and exposure to these structural impediments. It further addresses difficulties that governments, businesses and civil societies face with a focus on solving them. Lastly, it anticipates a paradigm shift away from the GDP growth-based, fossil fuel-driven industrial type of economic development towards a more inclusive and equitable model comprising eco-efficient low-carbon enterprises and economies. The chapter concludes that only the equitable, more inclusive and democratic developmental regimes are capable of consolidating sustainable development.

1.3.2 Part II: Strategies to Achieve Sustainable Economy and Zero Emissions Future

Integrating emissions reduction into planning and environmental laws is a crucial mechanism in helping subnational states or provinces play a role in mitigation measures. In states like New South Wales (NSW), Australia, the absence of an integrated approach to considering and reducing greenhouse gas emissions is a critical law and policy gap that needs to be addressed. Chapter 5 focuses on reducing greenhouse gas emissions (mitigation) and sets out how planning and development control systems can be part of the solution to achieving emissions reduction targets. It highlights two major structural barriers to taking effective action to reduce emissions in NSW, Australia's most populous state. The first is a lack of legislated or binding greenhouse gas emissions reduction targets supported by regulatory infrastructure or agency responsibility for reducing emissions. The second is a lack of integration between the need to reduce greenhouse gas emissions and the land-use planning system. The chapter focuses on the *Environmental Planning and Assessment Act 1979 (NSW)* because most greenhouse gas emissions from this state are authorised by planning and development approvals (explicitly or otherwise). It considers six key stages of the NSW planning system that are relevant to greenhouse gas emissions reduction. Within each of these key stages, the chapter illustrates how greenhouse gas emissions are currently dealt with and then how the law can be improved to help reduce emissions. Overall, the aim was to make clear that planning and development agencies and decision makers need stronger laws and guidance in achieving emissions control. It concludes with 14 recommendations to address this problem.

The transportation industry can be regarded as a significant contributor to greenhouse gases, which puts pressure on global logistics companies to disclose their carbon emissions in the form of carbon reports. According to Chap. 6, the majority of these reports show differences in the measurement and in reporting of carbon information. Using an institutional theory lens, this chapter discusses the emergence and the institutionalisation of carbon disclosure and their influence on the carbon reporting, using the two cases of FedEx and UPS. In particular, the chapter examines whether the carbon reports follow a symbolic or substantial approach and which institutional logic dominates the rationale behind each company's carbon disclosure. The chapter uncovers significant variances between those companies that provide insights into the different dominant logics and their influence on carbon reporting. The results show that both companies adopt different dominant logics due to heterogeneous carbon reporting practices. In a quest for market-controlled sustainability initiatives, this research is important: for understanding ways in which major carbon contributors operate between market and sustainability logics, and for policymakers to provide an understanding of how to encourage investments in carbon performance without reducing the ability to compete in the marketplace, thus initiating a shift to more environmentally friendly activities.

Chapter 7 reviews the treatment of population in climate change scenarios and the prospects for proactive interventions to influence outcomes. Sensitivity analyses have demonstrated population to be a dominant determinant of emissions. The assumption that population growth is determined by economic and educational settings is not well supported in historical evidence. Indeed, economic advance has rarely been sustained where fertility remained above three children per woman. In contrast, population-focused voluntary family planning programmes have achieved rapid fertility decline, even in very poor communities, and enabled more rapid economic advance.

Policy-based projections of global population have been presented, based on the historical course of nations that implemented effective voluntary family planning programmes in the past. If remaining high-fertility nations adopted such programmes, global population could yet peak below 9 billion by 2050. Current trends show that it is more likely to exceed 13 billion people by 2100, unless regional population pressures cause catastrophic mortality rates from conflict and famine. Global support for family planning could reduce population by 15% in 2050 and 45% in 2100 compared with the current trend. Co-benefits include gender equity, child health and nutrition, economic advancement, environmental protection and conflict avoidance.

1.3.3 Part III: Major Challenges Towards a Sustainable Future: Case Studies from the Asia-Pacific

Chapter 8 essentially is exploratory in nature. The changing terms of political legitimacy and the economic imperative to develop high-tech industry and the service sector generally provide clear motivation for the government to embrace an aggressive climate change strategy in China. These dual motivations have been manifested in China's change in international actions as evidenced by the very different outcomes of COP15 and COP21. The Chinese Government now has compelling and logical reasons to continue to take the initiative in setting ambitious environmental targets both domestically and with its international partners. This raises the question of how such momentum towards renewable energy sources will impact China's trading partners. Countries such as Australia, that export commodities to China, will need to be responsive to potential changes in market conditions. Indeed, China's policy in the area of renewable energy may have substantial implications for the approach taken by other countries in the region. Further, China's growing geopolitical influence may extend such implications to countries in other regions, particularly developing countries that are in any way tied to China politically or economically. The chapter was not intended as a definitive assessment of China's climate change or renewable energy policies. There exists, however, great scope for substantive quantitative and qualitative research to better understand China's evolving position with regard to both climate change and the unprecedented scale of the shift towards renewable energy. Indeed, those trading partners that best understand the motivations and economic underpinnings of China's shift to renewable energy will be best placed to pre-empt and respond to resultant changes in market demands.

According to Chapter 9, the emission intensity of India's GDP declined by more than 30% during the period 1994–2007 due to the efforts and policies that India has been following. India wants to further reduce the emission intensity of GDP by 20–25% between 2005 and 2020 by following the path of inclusive growth. But the surprise demonetisation of the high-value notes by the Prime Minister of India in 2016 has caused a total imbalance in the economy. According to Indian economists, the GDP growth rate will fall by 2% and the inflation rate will also remain high in the short to medium term. In this situation, it will be difficult for the country to satisfy its commitments in emissions control. The global scenario is also changing and will have great impact on funding the INDC. If the funding of the projects becomes uncertain, then their implementation may not materialise. However, this chapter hopes, in the long run, that the effect of demonetisation will be good and that if the economy recovers it will be in a position to satisfy its commitments on the INDC to the world.

It is widely accepted that human mobility caused by environmental change will take place internally first within affected countries rather than across borders. Chapter 10 examines the link between environmental vulnerabilities and human migration in various socioeconomic contexts. Previous studies have examined population mobility in response to vulnerability driven by sudden natural hazards like cyclone, flood and earthquake. However, little is known about the dynamics of human mobility in response to slow-onset hazards like drought. This study is based on comprehensive fieldwork with socioeconomically disadvantaged migrants who are exposed to seasonal drought in northern rural areas of Bangladesh. The study focused on a better understanding of how the affected individuals and families make decisions to either stay or to migrate internally in response to seasonal drought and other socioeconomic vulnerabilities. By adopting a case study approach, rural-to-urban migrants and their family members in the northern highland area of the country known as the Barind Tract were interviewed. The results suggest that migration decisions are consolidated by a variety of stressors including environmental and nonenvironmental components. The research found that some interventions implemented by government and nongovernment organisations are posing long-lasting impacts on the sustainability of rural livelihood, with a propensity to increase not reduce outward migration. These interventions have been debated and recommendations are made to address this emerging and complex livelihood problem in the context of adaptation issues due to global warming-induced climate change.

Chapter 11 of this volume explores the fiscal incentives and disincentives that contribute either positively or negatively to reducing emissions from deforestation and forest degradation (REDD+) in Indonesia. Indonesia is an important participant in the UNFCCC programme on REDD+. The programme is funded through financial contributions from developed to developing countries, which can eventually be part of a country's nationally determined contribution to reducing emissions, either domestically or via international emissions trading. The study finds that there are a number of formal charges, fees and taxes that apply to forest-related activities in Indonesia, which are stipulated within regulations promulgated by various government departments. A range of informal subnational charges also apply to forest-related activities, which has often provided a monetary incentive for local government, especially forest-rich districts, to exploit their timber resources. However, this has been proven as a disincentive for REDD+ implementation in Indonesia. The study also finds that there is a need for improved financial governance in future fiscal policy reform, which should include the removal of perverse incentives for forest conversion, the equitable and accountable distribution of financial incentives, the prevention of corruption and fraud, and the strengthening of economic benefits for smallholders. The study has recommended that in implementing the REDD+, the Government of Indonesia should consider providing incentives for the nonexploitation of forests by businesses engaged in the provision of environmental services as well as carbon transactions. This could take the form of private investments, private–public partnerships or civil society engagement in forestry and land-use change, and may include incentives such as payment for ecosystem services and for forest ecosystem restoration.

The concluding chapter articulates a method for investigating carbon emissions based on a carbon budget approach. The steps are:

- Amend the stocktake mechanism of Paris Agreement to include a carbon emissions cap whereby the global carbon emissions budget of 800 billion tonnes of CO_2e can be negotiated by Parties between 2020 and 2100.
- Each country determines their NDCs in line with their budgeted carbon emissions based upon equity and common but differentiated responsibilities and respective capabilities.
- Publish INDCs that conform to the global carbon emissions budget approach as a process of negotiation prior to stocktake mechanism review. Not only are INDCs published but expert review by the UNFCCC plays a role in providing feedback on the effectiveness of achieving an emissions trajectory that conforms to a global carbon emissions budget.

As mentioned earlier, this volume is based on a conference held in Brisbane, Australia in November 2016 organised at Griffith University, 1 year after the Paris Agreement was drawn. The main objective of the conference was to identify the key issues that need to be addressed in order for nations to achieve their INDCs. The overall outcome of this volume suggests that there are considerable challenges for nations to achieve the goals of the Paris Agreement. As stated earlier, members agreed that "If greenhouse gas emissions are cut by 40–70% below current levels by 2050, warming can be kept below 2 °C at reasonable cost by 2100". In this volume, our efforts were to analyse how to bridge the gap between COP21 commitments and 2030 targets of emissions control while employing an interdisciplinary approach.

References

Berkes, F, Colding, J and Folke, C, (2003). *Navigating social-ecological systems: Building resilience for complexity and Change*. Cambridge: Cambridge University Press.

Bruun, O., & Casse, T. (Eds.). (2013). *On the frontiers of climate and environmental change: Vulnerabilities and adaptation in Central Vietnam*. Heidelberg: Springer.

Ensor, J. (2011). *Uncertain futures: Adapting development to a changing climate*. Rugby: Practical Action Publishing.

Gibson, R. (Ed.). (2016). *Sustainability assessment: Applications and operations*. London: Routledge.

Giddens, A. (2009). *The politics of climate change*. Cambridge: Polity Press.
Klein, N. (2015). *This changes everything: Capitalism vs. climate change*. London: Penguin.
Matson, P, Clark, C. W and Anderson, K, (2016). *Pursuing sustainability: A guide to the science and practice*. Princeton: Princeton University Press.
Pelling, M. (2011). *Adaptation to climate change—from resilience to transformation*. Abington: Routledge.

Part I
Critical Perspectives on Achieving a Zero Emissions Future

Chapter 2
Joining the Dots: Sustainability, Climate Change and Ecological Modernisation

Michael Howes

Abstract Ecological modernisation (EM) can provide an effective strategy for improving sustainability and addressing climate change by overcoming the resistance to change in key sectors. This chapter triangulates a broad range of sources from the academic literature and synthesises them with the findings of previous research by the author. It is argued that governments can assist with the transition to a more sustainable low-carbon economy by using EM to design policies that promote technological innovation, engage with economic imperatives, implement institutional change, improve community engagement and change the public discourse to focus on practical "win–win" scenarios. The theoretical framework of EM was developed in Europe during the 1980s and will need to be adapted for countries like Australia that have a very different political context. Examples are given of how business models can be changed and how government policies can encourage the transition. The adoption of this proposal would increase community empowerment and improve democratic decision-making. This chapter undertakes an original synthesis of sustainable development, climate change mitigation and strong EM to create practical changes to business models.

Keywords Sustainable development • Sustainability • Climate change mitigation • Ecological modernisation • Energy

2.1 Introduction

It has become a cliché to state that the world faces a profound set of interlinked social, economic and environmental problems. Consider the state of the global environment. For the last five decades, humanity's ecological footprint has exceeded the capacity of the planet to provide resources sustainably and continues to rise,

M. Howes (✉)
Griffith School of Environment, Griffith University, Southport, Australia

Cities Research Institute, Griffith University, Southport, Australia
e-mail: m.howes@griffith.edu.au

biodiversity has declined substantially and greenhouse gas emissions have almost doubled (Howes, 2017). These problems have major economic and social dimensions, yet have been growing despite concerted efforts to respond to them from the international to the local level of governance (Howes et al., 2017).

Sustainable development and climate change policies were supposed to address such problems, but progress has been stymied by domestic politics, particularly in countries like Australia and the USA whose economies rely on high energy and material throughputs (Howes, 2005, 2017). Australia, for example, has developed a large resources sector that makes a significant contribution to exports, jobs and tax revenues. This makes the sector a powerful influence on national and state government policymaking, particularly with regard to climate change (Pearse, 2009). Such influence is replicated in several developed and developing countries where the economy is prioritised over the environment (Howes et al., 2017).

At the heart of the resistance to change is the fear that actions that improve environmental quality will impair business and cost jobs. The idea of sustainable development was supposed to allay this fear by decoupling prosperity from environmental damage. It appears, however, that major political and economic decision makers remain unconvinced of the credibility of this "win–win" scenario (i.e. they do not believe that it is possible to consistently look after both the environment and the economy), nor do they seem to understand the urgency of environmental issues (Howes, 2005, 2017; Howes et al., 2010, 2017). This chapter argues that the idea of ecological modernisation could be used to break the impasse by encouraging technological change, constructively engaging with economic imperatives, reforming political institutions, improving community engagement and transforming the public discourse so that the focus is on finding practical win–win scenarios. The next section outlines the nature of the problem in terms of the disconnection between the commitments made on sustainability and climate change and the lack of on-the-ground action. The concept of ecological modernisation (EM) is then explained and applied to the problem of climate change and energy production.

2.2 The Problem

The idea of sustainable development has a long history. The link between environmental, social and economic issues was recognised in the United Nations International Development Strategy of 1970 and was advanced further at the 1972 Conference on the Human Environment in Stockholm (Howes, 2005; Howes et al., 2017). The term "sustainable development" was coined in the final section of the 1980 World Conservation Strategy which suggested that humanity could have both a healthy economy and a healthy environment if it was more careful about how natural resources were used (IUCN, UNEP and WWF, 1980). This idea was picked up by governments as a way to resolve the growing tensions between environmentalists and business.

The World Commission on Environment and Development was established in 1983 and after four years of investigation its final report, Our Common Future (WCED, 1987), offered three useful outcomes. First, it gave a systematic catalogue of interlinked environmental, social and economic problems. Second, it offered sustainable development as a solution to the problems which is defined as: "development that meets the needs of the present without compromising the ability of future generations to meet their own needs" (WCED, 1987, Chap. 2, Para 1). Third, the report offered a framework strategy for change which included a meeting of national governments to negotiate new international agreements and actions. This led to the 1992 Rio Earth Summit, where governments of the world agreed to pursue sustainable development via Agenda21 (a document that covered all major sectors of the economy). This was supported by a statement of sustainability principles as well as the United Nations Framework Convention on Climate Change (UNFCCC), the Biodiversity Convention and a statement on forest principles (Howes, 2000; Howes et al., 2017). Subsequent summits in 1997, 2002 and 2012 allowed governments to reaffirm their commitment and assess progress. This culminated in the adoption of 17 Sustainable Development Goals in 2015. Goal 13 requires governments to "take urgent action on climate change and its impacts" and climate change was cited in many other goals (UN, 2015).

The UNFCCC came into force in 1994 and covers both mitigation (by reducing greenhouse gas emissions) and adaptation. The first Conference of Parties (COP) was held in Berlin in 1995 and these annual COPs are the major forum for coordinating international action. The 1997 COP3, for example, led to the Kyoto Protocol that committed developed countries to reducing their emissions by 5% on their 1990 baseline average by 2012 as a first step. In Cancun, the 2012 COP15 saw signatories agree to limit the rise in average global temperatures to no more than 2 °C above pre-industrial levels in order to avoid the worst impacts of climate change. More recently, The 2015 COP21 in Paris set a more ambitious aspirational limit of 1.5 °C and agreed to pursue a "sustainable low carbon future" (UN, 2016). These developments have been driven by a series of assessment reports by the Intergovernmental Panel on Climate Change that have been released every 5 years since 1990 (IPCC, 2014).

The parallel streams of international policymaking on sustainability and climate change have several common features. First, research into the state of the planet has been pivotal in getting issues onto the policy agenda. Second, these issues require action by all sectors of society (the state, business and the community) at all levels (from the international to the local). Third, international negotiations have often occurred prior to the creation of domestic policies (Howes, 2005). Fourth, despite international commitments being in place for a quarter of a century, and agreements stating that these issues are urgent, there has been a lack of progress in implementing the on-the-ground changes that are needed to address climate change and improve sustainability (Howes et al., 2017). As a result, resources continue to be depleted, biodiversity continues to be lost, pollution remains a problem, the ecosystems on which we depend continue to be undermined, greenhouse gas emissions have risen and vulnerability to the impacts of climate change has increased (IPCC, 2014; UNEP, 2012; UN MEA, 2005; WWF, 2014). The Asia-Pacific region has seen rapid growth in primary and secondary industries over the last few decades which has contributed to these problems, and the region is also home to some of the most vulnerable populations (AAS, 2015; IPCC, 2014; WHO, 2015).

This disconnect between international commitments and local action can in large part be attributed to the barriers to action created by domestic politics (Howes, 2005; Howes et al., 2017). In Australia, for example, a significant proportion of the economy relies on the extraction and export of fossil fuels as well as energy-intensive industries such as metal production. Actions to reduce greenhouse gas emissions are seen as a threat to these industries that rely on the burning of fossil fuels (Howes et al., 2010). This has led to a concerted campaign by some industry leaders to stall climate change policies (Pearse, 2009) and encouraged an ongoing debate about whether climate change is happening. Some members of parliament openly deny the overwhelming scientific evidence.[1] In such an environment, both major political parties have found it difficult to take effective action. Even when some progress has been made, it is often undone by major policy reversals at the national, state or local level of government (Howes & Dedekorkut-Howes, 2016).

The underlying fear of some leaders in business, the state and the community is that taking action on climate change, particularly with regard to reducing greenhouse gas emissions, will cost jobs, impose extra costs on businesses that will make them uncompetitive internationally, and reduce economic growth. This chapter argues that this fear can be allayed by developing a transition strategy based on the concept of strong EM that can improve sustainability.

2.3 Ecological Modernisation

Ecological Modernisation (EM) is a school of thought that approaches the issue of sustainability as a design challenge. The aim is to decouple economic prosperity from environmental damage (Howes et al., 2010). This can be achieved by transforming the key technical, economic, social and political institutions of modernity to encourage the uptake of more eco-efficient production and consumption. Such a move should reduce resource use, waste and pollution and reduce the pressures being put on the state of the environment (Berger, Flynn, Hines, & Johns, 2001; Christoff, 1996; Dryzek, 2005; Howes, 2005; Huber, 2008; Janicke, 2008; Janicke & Jacob, 2004; Mol & Spaargaren, 2000). This transformation is good for the environment and good for business because more eco-efficient firms spend less money on inputs such as raw materials and energy (Curran, 2009; Gouldson & Murphy,

---

[1] An excellent example of this occurred on 15 August 2016 on the television programme Q&A, which screened on the Australian Broadcasting Corporation. During the program, Professor Brian Cox (a physicist) confronted new Senator-elect Malcom Roberts with the evidence of climate change. Roberts refused to believe the data provided and asserted that it had been manipulated by NASA so that he could continue to deny that climate change was happening. You can find the programme transcript at http://www.abc.net.au/tv/qanda/txt/s4499754.htm.

1997; Janicke, 2008). According to this school of thought, governments should create policies that encourage and support this transition (Blowers, 1997; Huber, 2008; Janicke, 2008; Janicke & Jacob, 2004; Lundqvist, 2000; Mol & Sonnenfeld, 2000; Weale, 1998). This is what governments are supposedly attempting to do with their sustainability policies (Howes, 2005; Howes et al., 2010).

EM emerged in parallel to the idea of sustainable development in the 1980s from European scholars such as Martin Janicke and Joseph Huber (Grant & Papadakis, 2004; Hajer, 1995; Huber, 2000, 2008; Janicke, 2008; Janicke & Jacob, 2004; Mol & Sonnenfeld, 2000; Mol & Spaargaren, 2000; Weale, 1998). The early versions of EM focused on technological innovation and were criticised for adopting a techno-corporatist approach to social change (Christoff, 1996). As the school developed, stronger versions emerged that expanded the focus to embrace substantial institutional changes in political and economic systems. The idea was to both create incentives for change and provide better feedback mechanisms for democratic decision-making (Christoff, 1996; Dryzek, 2005; Fisher & Freudenburg, 2001). Strong EM has much in common with the broader goals of sustainable development policies and has five core themes for change (Howes, 2005; Howes et al., 2010):

1. Technological innovation to increase the efficiency of resource use while reducing waste and pollution
2. Engaging with economic imperatives in order to create ongoing financial incentives for reducing environmental impacts
3. Political and institutional change that moves from a top-down model of government to more collaborative governance which builds partnerships across sectors
4. Transforming the role of social movements in order to revitalise civil society and its ability to provide a more effective feedback to decision makers
5. Discursive change that addresses fears about "jobs versus the environment" by reframing the issue as an opportunity to pursue win–win scenarios (i.e. solutions that are good for business and the environment)

A range of similar ideas have cropped up in different parts of the world. Cleaner production and natural capitalism, for example, is an idea that underpins many environmental management systems, such as ISO 14000, the Eco-Management and Audit Scheme and Pollution Prevention Pays (Von Weizsacker, Hargroves, Smith, Desha, & Stasinopoulos, 2009). All of these involve analysing material and energy flows at the facility level in order to make changes that reduce both wastage and costs (Howes, 2005). At the next level is the idea of industrial ecology, where different facilities are linked so that the waste from one becomes the input of the next (much like an ecosystem). There are also the ideas of life-cycle analysis, cradle-to-cradle design, biomimicry and green design that encourage a rethink of products and processes as well as consumption and disposal. While these ideas and initiatives have often grown up in isolation, they all approach sustainability as a design challenge and promote eco-efficiency. Hence, they can be grouped under the broad umbrella of strong EM that provides the theoretical underpinning for the idea of sustainability (Howes et al., 2010).

A practical example of EM at the firm level is the US-based carpet manufacturer Interface which embraced biomimicry in order to become more sustainable (Suzuki

& Dressel, 2002). In 1996, the CEO, Ray Anderson, decided to aim to make the company sustainable by 2020. Since then it has eliminated toxic substances from the manufacturing process, switched inputs from petrochemicals to natural fibres, redesigned products to be recyclable, reduced the intensity of water use by 87%, cut solid waste and increased material recycling. With regard to climate change, Interface now gets 84% of its energy from renewable sources, it has increased energy efficiency by 45% and greenhouse gas emissions per unit of production have been cut by 92% (Interface, 2016a). The company is a successful business: it has expanded its operations to 110 countries, income has grown to more than US$1 billion, gross profit on sales is more than US$382 million and the share price has risen by 26% in the last 5 years (Interface, 2016b). This dramatically demonstrates the win–win scenario where institutional and technological change can improve both the economic and environmental performance of an organisation, and there are many others (see: Howes, 2005; Suzuki & Dressel, 2002; Von Weizsacker et al., 2009).

2.4 EM, Climate Change and Energy Production

Climate change is one of the key threats to sustainability and could be addressed using a strategy derived from the core themes of strong EM. The main driver of climate change is anthropogenic greenhouse gas emissions and most of these come from the burning of fossil fuels by the energy and industrial sectors (AAS, 2015; IPCC, 2014). Consider the life-cycle impacts of burning coal to generate power. First, extracting coal from an open-cut mine has several impacts: the mining process depletes a non-renewable resource; preparing the site requires vegetation to be cleared which would otherwise absorb carbon dioxide and provide a habitat for native species (hence reducing biodiversity); clearing also increases soil erosion and land degradation; and runoff from the site can pollute local waterways. Second, transporting the coal to a power station releases greenhouse gases from the fuel burnt in the vehicles or vessels involved. Third, burning the coal in the power station releases more greenhouse gases and other air pollutants (such as particulates, oxides of sulphur and nitrogen, and mercury) that put further pressure on the state of the environment and human health (WHO, 2015). Industries such as manufacturing and metal smelting require large amounts of energy and increase the demand for coal extraction and use. The net greenhouse gas emissions of this cycle drive climate change which puts further pressure on biodiversity as well as having significant social and economic costs. This is a classic example of market failure due to a negative externality where some of the costs of production are imposed on people who are not involved in the transaction (i.e. vulnerable residents of poorer countries with low per capita greenhouse gas emissions and future generations) (Garnaut, 2008; Stern, 2006). Applying the five core themes of EM can help to correct this market failure and push the sector towards sustainability.

First is the need to shift to technology that cuts greenhouse gas emissions from power generation. Fortunately, there are already several viable renewable energy options that have significantly fewer environmental impacts than coal-fired electricity generation, including wind, solar, geothermal and solar thermal. Second, there needs to be consistent economic incentives to encourage the uptake of this technology. There have been some trials of subsidies and feed-in tariffs for rooftop solar, while various renewable energy targets have been set at the national and state level. There will need to be a price placed on greenhouse gas emissions in order to internalise the negative externality that they represent. This could be in the form of a carbon tax, a cap and trade scheme, or a baseline and credit scheme. The politics of this are difficult, particularly given recent major policy reversals. This is where the third theme of political institutional change may assist, particularly with regard to building partnerships with the renewable energy sector to promote projects in regional areas. On the fourth theme of engaging with social movements, the Ecologically Sustainable Development working groups provide a useful example of how this might be achieved. They were run by the Australian Government from 1990 to 1992 and included representatives from all levels of government, business, unions and environmental groups. The outcome was the 1992 National Strategy for Ecologically Sustainable Development that still has bipartisan political support and is enshrined in legislation at the national and state levels of government (Howes, 2005). Finally, and perhaps most importantly, is the necessary discursive change. By promoting the win–win scenario and real-world case studies such as Interface, the public debate could be shifted away from the "jobs versus the environment" false dilemma and business could be encouraged to see opportunities for improving profitability. Obviously, this is going to be difficult in Australia as the economy has a significant primary industries sector based on fossil fuels and energy-intensive industries, but it is not impossible (Howes, 2015; Howes et al., 2010).

Consider one of the hardest cases: a firm that is currently in business to extract and sell thermal coal to be burnt for energy production. How could such a business be ecologically modernised to be more sustainable and reduce greenhouse gas emissions? As in the case of Interface, such a transformation will require a commitment from the top executives to rethink the business model. This must occur on the face of some major challenges for the industry. The price that Australia receives for its coal has fallen significantly over several years with demand starting to fall in the OECD and China (World Bank, 2016a). Further, at the 2015 COP21 in Paris, 195 governments committed themselves to reducing greenhouse gas emissions, more than 90 of the Intended Nationally Determined Contributions submitted included emissions trading schemes and more than 40 countries have already imposed a price on carbon (World Bank, 2016b). In the absence of viable large-scale carbon capture and storage, these developments suggest that the future of demand for thermal coal is not promising. A wise strategy for a firm that is dependent on revenue from thermal coal production would therefore be to diversify the business into other activities that are profitable and growing, and which build on the company's strengths. Coal businesses understand the energy sector, have a great deal of geological expertise, are experienced with heavy engineering and are able to operate in remote areas. This makes them ideally placed to enter the geothermal energy sector. As the demand for renewable energy grows, they can shift capital and employees out of coal mining and into the construction and operation of geothermal power stations. Governments could assist by providing policy consistency in the form of well-designed economic incentives and by entering into public–private partnerships to provide the necessary infrastructure. Clever firms could also constructively engage with community organisations to highlight their improving environmental performance and build social capital. All of this could be done if business, political and community leaders genuinely believed that such a transition would deliver better economic, social and environmental outcomes (Howes, 2005; Howes et al., 2010, 2017).

2.5 Conclusion

The global environment has been seriously degraded over many decades, and this has significant social and economic ramifications that spill into the political arena. Climate change is one of the more critical problems generated by this degradation. It is clear that the international commitments to pursue sustainable development and address climate change need to be matched by on-the-ground changes to the way things are produced and consumed. This is particularly important when it comes to the energy sector that underpins all aspects of industrialised societies. Economic prosperity needs to be decoupled from greenhouse gas emissions, and this means a switch to renewable energy on a global scale. This transition, however, is resisted by sectors that are built upon the extraction and sale of fossil fuels, like thermal coal, because it poses a threat to the prevailing business model. In order to survive, this sector will need to rethink its business model and diversify into renewable energy. Governments can help by providing policies that encourage the necessary technological innovation and create positive economic incentives for change. They will also have to be prepared to reform governing institutions in order to promote more collaboration and partnerships, as well as providing a constructive avenue for community engagement. Above all, fears that being good to the environment is bad for business need to be allayed by promoting practical win–win scenarios. These are the kinds of changes promoted by strong EM, which is why it can provide a practical transition strategy. It must be noted, however, that EM is not a panacea for all the world's social, economic and environmental problems. It is simply a useful device that can help to overcome the resistance to change in some key sectors.

References

Australian Academy of Science [AAS]. (2015). *The science of climate change: Questions and answers*. Canberra: .AAS. Retrieved from https://www.science.org.au/learning/general-audience/science-booklets-0/science-climate-change

Berger, G., Flynn, A., Hines, F., & Johns, R. (2001). Ecological modernization as a basis for environmental policy: Current environmental discourse and policy and the implications on environmental supply chain management. *Innovation, 14*(1), 55–72.

Blowers, A. (1997). Environmental policy: Ecological modernisation or risk society? *Urban Studies (Edinburgh, Scotland), 34*(5-6), 845–871.

Christoff, P. (1996). Ecological modernisation, ecological modernities. *Environmental Politics, 5*(3), 476–500.

Curran, G. (2009). Ecological modernisation and climate change in Australia. *Environmental Politics, 18*(2), 201–217.

Dryzek, J. (2005). *The politics of the Earth: environmental discourses*. Oxford: Oxford University Press.

Fisher, D., & Freudenburg, W. (2001). Ecological modernisation and its critics: Assessing the past and looking toward the future. *Society & Natural Resources, 14*, 701–709.

Garnaut, R. (2008) *The Garnaut climate change review*. Melbourne: Cambridge University Press. Retrieved from http://www.garnautreview.org.au/2008-review.html

Gouldson, A., & Murphy, J. (1997). Ecological modernization: Restructuring industrial economics. In M. Jacobs (Ed.), *Greening the millennium? The new politics of the environment* (pp. 74–86). Oxford: Blackwell Publishers.

Grant, R., & Papadakis, E. (2004). Challenges for global environmental diplomacy in Australia and the European union. *Australian Journal of International Affairs, 58*(2), 279–292.

Hajer, M. (1995). *The politics of environmental discourse: Ecological modernization and the policy process*. Oxford: Clarendon Press.

Howes, M. (2000). A brief history of Commonwealth sustainable development policy discourse. *Policy, Organisation and Society, 19*(1), 65–85.

Howes, M. (2005). *Politics and the environment: Risk and the role of government and industry*. London: Earthscan.

Howes, M. (2015). Australia's 'climate roundtable' could unite old foes and end the carbon deadlock. *The Conversation*. Retrieved from https://theconversation.com/australias-climate-roundtable-could-unite-old-foes-and-end-the-carbon-deadlock-44007

Howes, M. (2017). After 25 years of trying, why aren't we environmentally sustainable yet?" *The Conversation*. Retrieved July 12, 2017 from https://theconversation.com/after-25-years-of-trying-why-arent-we-environmentally-sustainable-yet-73911

Howes, M., & Dedekorkut-Howes, A. (2016). The rise and fall of climate adaptation governance on the Gold Coast, Australia. In J. Knieling (Ed.), *Climate adaptation governance: Theory, concepts and Praxis in cities and regions* (pp. 237–250). Hamburg: Wiley.

Howes, M., McKenzie, M., Gleeson, B., Gray, R., Byrne, J., & Daniels, P. (2010). Adapting the idea of ecological modernisation to the Australian context. *Journal of Integrative Environmental Sciences, 7*(1), 5–22.

Howes, M., Wortley, L., Potts, R., Dedekorkut-Howes, A., Serrao-Neumann, S., Davidson, J., … Nunn, P. (2017). Environmental sustainability: A case of policy implementation failure? *Sustainability, 9*(2), 165. Retrieved July 12, 2017 from http://www.mdpi.com/2071-1050/9/2/165/htm

Huber, J. (2000). Towards industrial ecology: Sustainable development as a concept of ecological modernization. *Journal of Environmental Policy and Planning, 2*, 269–285.

Huber, J. (2008). Pioneering countries and the global diffusion of environmental innovations: These from the viewpoint of ecological modernisation theory. *Global Environmental Change, 18*, 360–367.

Interface. (2016a). *Environmental footprint: All metrics*. Atlanta, USA. Retrieved from http://www.interfaceglobal.com/Sustainability/Our-Progress/AllMetrics.aspx

Interface. (2016b). *Annual Report 2015*. Atlanta, USA. Retrieved from http://www.interfaceglobal.com/Investor-Relations/Annual-Reports.aspx

Intergovernmental Panel on Climate Change [IPCC]. (2014). *Climate change 2014: Synthesis Report*. IPCC. Retrieved from http://www.ipcc.ch/report/ar5/syr/

IUCN, WWF & UNEP [International Union for the Conservation of Nature and Natural Resources, the World Wildlife Fund and the United Nations Environment Programme]. (1980). *World conservation strategy*. Geneva: UN Publications.

Janicke, M. (2008). Ecological modernisation: New perspectives. *Journal of Cleaner Production, 16*, 557–565.

Janicke, M., & Jacob, K. (2004). Lead markets for environmental innovations: A new role for the nation state. *Global Environmental Politics, 4*(1), 29–46.

Lundqvist, L. (2000). Capacity-building or social construction? Explaining Sweden's shift towards ecological modernisation. *Geoforum, 31*, 21–32.

Mol, A., & Sonnenfeld, D. (2000). Ecological modernisation around the world: An introduction. *Environmental Politics, 9*(1), 3–14.

Mol, A., & Spaargaren, G. (2000). Ecological Modernisation. *Environmental Politics, 9*(1), 17–49.

Pearse, G. (2009). Quarry Vision: Coal, climate change, and the end of the resources boom. *Quarterly Essay, 33*, 1–122.

Stern, N. (2006). *Stern review on the economics of climate change*. London: HM Treasury/Cabinet Office.

Suzuki, D., & Dressel, H. (2002). *Good news for a change: Hope for a troubled planet*. Sydney: Allen & Unwin.

UNEP (United Nations Environment Programme). (2012). *Global Environmental Outlook Report 5*. Retrieved from http://www.unep.org/geo/geo5.asp

United Nations [UN]. (2015). *Sustainable development goals*. Retrieved from http://www.un.org/sustainabledevelopment/sustainable-development-goals/

United Nations [UN]. (2016). *Framework convention on climate change*. Retrieved from http://unfccc.int/2860.php

UNMEA (United Nations Millennium Ecosystem Assessment). (2005). *Ecosystems and human well-being: Synthesis*. Washington, DC: Island Press. Retrieved from http://www.millenniumassessment.org/en/Synthesis.html.

Von Weizsacker, E., Hargroves, K., Smith, M., Desha, C., & Stasinopoulos, P. (2009). *Factor Five: Transforming the global economy through 80% improvements in resource productivity*. London: Earthscan.

WCED (World Commission on Environment and Development). (1987). *Our common future*. Melbourne: Oxford University Press.

Weale, A. (1998). The politics of ecological modernization. In J. Dryzek & D. Schlosberg (Eds.), *Debating the Earth: The environmental politics reader* (pp. 301–318). Oxford: Oxford University Press.

WHO (World Health Organisation). (2015). *Air Pollution*. Retrieved from http://www.who.int/topics/air_pollution/en/

World Bank. (2016a). *Commodity Prices Outlook: July 2016*. Retrieved from http://www.worldbank.org/en/research/commodity-markets

World Bank. (2016b). *Carbon pricing watch 2016*. Retrieved from https://openknowledge.worldbank.org/handle/10986/24288

WWF (World Wide Fund for Nature). (2014). *Living Planet Report 2014*. Retrieved from http://www.wwf.org.au/our_work/people_and_the_environment/human_footprint/living_planet_report_2014/

Chapter 3
A Systems Critique of the 2015 Paris Agreement on Climate

Luke Kemp

Abstract This chapter presents a systems dynamics critique of the 2015 Paris Agreement on Climate established at the 21st Conference of the Parties (COP21) to the United Nations Framework Convention on Climate Change (UNFCCC). Pledges under the Agreement are currently inadequate to limit global warming to safe levels. However, it is intended to be a system that evolves over time and changes state behaviour to encourage increasingly ambitious emissions reductions from members. The Agreement is designed to increase action through a ratchet mechanism obligating countries to put forward stronger targets, political pressure and a "signal" to investors to transition towards low-carbon portfolios and activities. An analysis using causal loop diagrams finds that none of these mechanisms for change are convincing. The legal wording of the Paris Agreement means that no "ratchet mechanism" exists. Political pressure through a pledge and review process has rarely worked in other international agreements or in previous international efforts on climate change. The idea of international law sending an investment signal is tenuous at best, and existing evidence in renewable energy and fossil fuel markets suggests that the signal is currently not functioning. Moreover, the Paris Agreement has inbuilt delay, making the lock-in of emissions-intensive trajectories likely. In short, the mechanisms for change designed into the Paris Agreement are unlikely to work. The Agreement as a system for changing state behaviour in a sufficient timescale is likely to fail.

Keywords Climate policy • COP21 • Paris • International relations • Environmental policy • Systems dynamics

L. Kemp (✉)
Fenner School of Environment and Society, Australian National University, Canberra, ACT, Australia

Crawford School of Public Policy, Australian National University, Acton, ACT, Australia
e-mail: luke.kemp@anu.edu.au

3.1 Introduction

International agreements are not static: the success or failure of any given agreement cannot be determined at a single point in time. Agreements are dynamic systems intended to drive or constrain change among states. The success of an international agreement depends on the credibility of its mechanisms to amplify or tame certain behaviours.

The Paris Climate Agreement of 2015 is one agreement that has widely been heralded as a diplomatic success (Harvey, 2015; Rajamani, 2016). But its success in addressing climate change is contingent on how it evolves over time. In this chapter, I will dissect the Paris Agreement as a system of feedbacks. In doing so, I will highlight how the system that Paris established is intended to succeed and why it will probably fail.

The Paris Agreement was adopted by consensus on the 12 December 2015. The Agreement is a treaty under international law and is binding upon member parties. It entered force and became legally binding on ratifying countries on 4 November 2016. Currently, 143 countries have ratified the Agreement (UNFCCC, 2017).

The overarching goals of the Agreement, stated under Article 2, are to limit "increase in the global average temperature to well below 2 °C above pre-industrial levels and to pursue efforts to limit the temperature increase to 1.5 °C above pre-industrial levels". This is later elaborated on in Article 4 which aims to achieve "balance between anthropogenic emissions by sources and removals by sinks of greenhouse gases in the second half of this century". There is some slight dissonance between the two aims: limiting to 2 °C, let alone 1.5 °C, will likely require net zero emissions closer to mid-century (Rockström et al., 2017).

To meet these ambitious long-term goals, the Paris Agreement establishes a pledge and review process that requires countries to put forward pledges every 5 years, with correlated reviews called "global stocktakes". The pledges for national climate action or nationally determined contributions (NDCs) under the Agreement are not legally binding. They are not "contracts" under international law, but political commitments (Bodansky, 2015).

The pledge and review cycles of Paris are depicted in Table 3.1. The first submission of new pledges will occur in 2025 and will be proceeded by the first global stocktake in 2023. These stocktakes will account for and review the cumulative efforts of countries under the Agreement and estimate the temperature rise and greenhouse gas concentration the world is tracking towards. It will also review collective progress in terms of international financing of climate efforts and adaptation activities. The next stocktake will then occur in 2028, followed by a new round of pledges in 2030. The cycle will continue ad infinitum until net zero emissions (or presumably ecological collapse) is reached.

The pledges put forward under the Paris Agreement are currently insufficient to limit global warming to 2 °C or 1.5 °C above pre-industrial levels (Rogelj et al., 2016). Instead, the cumulative effort of the pledges would likely lead to a temperature rise of 2.5–2.8 °C (CAT, 2017). Current policies would see a likely temperature rise of 3.3–3.9 °C (CAT, 2017). These are likely to be conservative estimates since the modelling they are based on does not include feedbacks in the climate system, such as the release of permafrost (Hatfield-Dodds, 2013).

3 A Systems Critique of the 2015 Paris Agreement on Climate

Table 3.1 The Paris Agreement process

	2015	2016	2017	2018	2019	2020	2023	2025	2028	2030
Process	Paris Agreement adopted	Open for signature and ratification Entry into force on 4 November		Paris Rulebook adoption		☆ Submission of 2030 pledges for parties only with 2025 pledges. Those with 2030 targets must "communicate or update" their pledges. ★ New round of pledges due for all parties.				
Review	Synthesis report		Synthesis report update	Facilitative dialogue.			Global stocktake		Global stocktake	
Pledges	Submission of intended nationally determined contributions					☆		★		★

The current inadequacy of climate pledges is a challenge, but not a fatal flaw. Paris is intended to be a dynamic instrument. It is by nature a "broad and shallow" international agreement that is envisioned to deepen over time (Aldy, Barrett, & Stavins, 2003). The deepening in this case is a strengthening of climate action by participating countries to bridge this ambition gap between long-term goal and current efforts. Many commentators have suggested that Paris will encourage compliance and increasing action through a "ratcheting" mechanism (Bodansky, 2016).

3.2 There Is No Ratchet

The pledge and review system of the Paris Climate Agreement is intended to hinge on the use of a ratchet mechanism: a measure that forces countries to periodically ratchet up their commitments. The problem is that, despite popular opinion, the ratchet mechanism does not exist.

This ratchet mechanism is supposed to exist across several provisions. Both the use of global stocktakes and the stipulation that future pledges be informed by the stocktake are part of the ratcheting mechanism. The stocktakes themselves, as noted later in this chapter, are inherently weak, while the need for future pledges to be "informed" by stocktakes in legal practice means next to nothing.

The main mechanism for ratcheting is situated under Article 4.3 of the Agreement which reads "Each Party's successive nationally determined contribution will represent a progression beyond the Party's then current nationally determined contribution…"

(UNFCCC, 2015). Unfortunately, the legal wording here is unhelpfully ambiguous. "Progression" is not defined and could be interpreted to justify even minor adjustments. Moreover, progression is for NDCs as a whole and is not specified for mitigation. There is a legal argument that a small increase in international climate financing accompanied by an unchanged mitigation target could be put forward as a progression. A change of baseline might also be used to frame a target as progressively stronger, while committing to weaker domestic action.

Even if progression in emissions cuts was required, it would be an impotent tool. An increase in terms of mitigation commitments could be entirely insignificant but still qualify as a progression. For example, the Australian pledge is to reduce emissions 26–28% below 2005 levels by 2030 (Australian Government, 2015). A progression on this target could be 29–30% by 2035. It would not be a progression in terms of the rate of emissions reductions but would be an increase in absolute terms. Alternatively, any country could put forward incremental progressions but never achieve them. Pledges under Paris are unbinding and there are no mechanisms to effectively enforce the fulfilment of legal obligations.

Article 4.3 may even create perverse incentives for weaker targets. If a country believes it needs to scale-up action over time, there may be an enticement to initially put forward weaker targets. That would make both the achievement of targets and progression easier. It is politically far safer for countries to put forward weaker targets and then overachieve rather than risk more stringent goals they may not meet (Raustiala, 2005). Article 4.3 and the illusion of a ratchet mechanism may just make this strategy more enticing for risk-adverse states.

Article 4.3 is not a dynamic instrument to increase action within the system of the Paris Agreement. It is a vague and unenforceable political expectation. It may have some impact, but it does not provide the stimulus necessary to bridge the gap between current efforts and what is needed to meet the 2 °C challenge.

3.3 Feedback to Success

While there is no effective legal ratchet mechanism, Paris does contain two feedback mechanisms that could bolster ambition over time.

The first is simple political pressure. The entire premise of a pledge and review system is that pressure from within countries and from other states can hold governments accountable to their targets (Bodansky, 2016). This is based on the belief that a pledge can be just as effective as a legal contract if the state believes its reputation and popular support will suffer if it publicly fails to meet it (Jacquet & Jamieson, 2016). The regular occurrence of a global stocktake every 5 years is intended to create political moments that can pressure governments into increasing ambition.

The second is the idea of a "market message" or "investment signal". This is the notion that long-term goals supported by pledges showcase to investors that the world is heading towards decarbonisation (van Asselt, 2016). This political agreement on the direction of travel creates regulatory certainty in low-emissions developments and uncertainty for emissions-intensive investments. This was succinctly expressed in a post-Paris op-ed in the Economist: "Perhaps the most significant effect of the Paris agreement in the next few years will be the signal it sends to investors ... after Paris, the belief that governments are going to stay the course on their stated green strategies will feel a bit better founded—and the idea of investing in a coal mine will seem more risky" (The Economist, 2015).

Figure 3.1 is a causal loop diagram of how these two feedback mechanisms should theoretically operate. In causal loop diagrams, arrows denote a flow of influence. The "variables" (boxed and unboxed labels) are key aspects of the system under investigation. Polarities (+ and –) show the nature of the relationship between variables. They are not moral judgements but simply highlight how one variable will change a connected variable. Positive signs (+) demonstrate that the direction of change of the variable at the tail of the arrow will be the same as that the head (e.g. increasing progress towards global goals increases the level of regulatory certainty). Negative polarities (–) signify the opposite: an increase in the variable at the tail of the arrow will decrease the variable at the head, and vice versa. The diagram in Fig. 3.1 is simplified to showcase the main causal loops, or feedback mechanisms, that are supposed to drive climate action under the Paris Agreement.

In Fig. 3.1, peer pressure acts as a balancing feedback loop (denoted by the "B"). A balancing feedback means that the system pushes back against the initial change to maintain equilibrium, much like a thermostat. If a country's efforts are insufficient then the review processes, such as the global stocktake, will politically pressure the government into strengthening their targets. Once global progress is on track to meet the goals under Article 2, there will presumably be significantly less political pressure for countries to further strengthen their efforts.

The investment signal is an amplifying or "reinforcing" feedback (represented by the "R" in the diagram). Reinforcing feedback loops are amplifying processes by which the initial change is driven even further by the system. These are often thought of as "snowball" or "runaway" effects. The presence of universally agreed global goals creates regulatory certainty that both deters emissions-intensive investments and promotes investments in low-emissions businesses and activities. The changing investment patterns lead to shifting business activities as the market adapts. This shift leads to decreases in the cost of renewable energy and other low-emissions technologies and ultimately drives further mitigation. As greater progress is made, more confidence and trust is placed in the credibility of these long-term goals. This leads to increased regulatory certainty, creating a virtuous cycle for emissions reductions.

Or at least it is supposed to. There are good reasons to doubt the effectiveness of both feedbacks.

3.4 The Pressure Valve

The Paris Agreement as a pledge and review system namely relies on political pressure for compliance and escalation. It is political expectation rather than legal obligation that is used to ensure that states meet their NDCs. This approach rests on the notion that countries will respond to the threat of loss of reputation. Missing targets

Fig. 3.1 The feedbacks of Paris

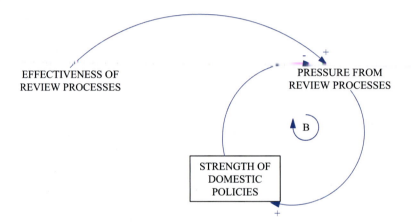

Fig. 3.2 The political pressure feedback

can lead to governments becoming pariahs on the international stage or experiencing a drop in domestic political support. The same political pressure should encourage countries to ratchet up their mitigation efforts over time. States that increase their targets sufficiently will be seen as good international players, or even leaders. Those who fall behind will be branded as laggards and free-riders.

Figure 3.2 provides a representation of this political pressure argument as a simple balancing feedback loop. Effective review processes create visible and credible evidence to base criticisms on. They also provide a high-profile political moment for countries, media and domestic groups to mobilise criticism. This pressure (hopefully) can encourage laggards to increase the strength of their domestic policies. As noted previously, as efforts become increasingly stronger, there should be commensurately less political pressure. This creates a balancing feedback loop. This is not the only balancing feedback in play. The impacts of climate change could also create heightening political pressure for stronger pledges and policies.

There are both empirical and instrumental reasons to suspect that political pressure cycles will not be effective, or at least effective enough. Empirically, international agreements that rely on political pressure have a mixed track record. Some agreements, such as the 1975 Helsinki Accords, the Geneva Conventions and World Health Organization's 1991 tuberculosis framework, have been relatively successful in deploying political pressure to push states to change and abide by their word (Heywood, 2013; Jacquet & Jamieson, 2016). However, many of these agreements have not been universally successful. While the Geneva Conventions have become a cornerstone of international humanitarian law, their precepts are still violated in international conflicts at an alarming frequency. It is also difficult to tell how much of the pressure has been created by the conventions themselves and how much is simply the power of the underlying norm of liberal humanism in the international system. There is no counterfactual to compare against, and it is therefore difficult to assess how effective political pressure has been and how much has been due to other underlying factors.

The previous examples of potential success are also qualitatively different to climate change. Abiding by the tuberculosis framework or the Geneva Conventions did not conflict with the profitability of existing incumbent industries or require any significant costs or significant changes to state behaviour. In contrast, meeting the goals of Paris requires a drastic economy-wide transformation (Rockström et al., 2017), a transformation that would result in large losses for the fossil fuel industry.

Previously, when such deep changes in state behaviour were required, pledges were not the instrument of choice. In the case of ozone-depleting substances, a more top-down approach using punitive non-party measures was taken. The World Trade Organization uses a judicial system through the dispute settlement mechanism whose verdicts are backed by trade measures. When structural changes are needed, the world has tended to opt for carrots and sticks with economic consequences underlined with legal authority. Simple political pressure has been left to deal with less pressing issues that require more shallow modifications to the international status quo.

Even if peer pressure were to work, the Paris Agreement is not properly designed to exert a large degree of pressure on individual states. The global stocktake and the transparency framework are not designed to publicly name and shame countries. The global stocktake is intended to review "collective progress" (UNFCCC, 2015). In negotiations since Paris, parties such as China and India have been adamant that the stocktake should avoid being a political or judgemental process, even at the collective level.[1] It should be a mechanical noting of efforts thus far. The transparency framework under Article 13 is to be applied in a manner that is "facilitative, non-intrusive, non-punitive manner, respectful of national sovereignty, and avoid placing undue burden on parties". Neither the transparency nor the global stocktake is likely to garner a large amount of public attention or provide significant ammunition for the reputational punishment of laggards.

In the context of Fig. 3.2, this means there is little reason to believe that the existing review mechanisms are effective enough to trigger a peer pressure causal loop. The efficacy of these mechanisms is dubious, particularly when viewed historically. They are by no means new.

---

[1] Author's personal observations of negotiations.

It is difficult to expect political pressure to ensure compliance and spur an energy revolution given the history of international climate policy. The climate negotiations have already been conducting a pledge and review system. The last pledge and review round was during 2015 and 2014, when almost all countries submitted intended NDCs (INDCs). These were then compiled and analysed by the UNFCCC secretariat in a "synthesis report" that was released shortly before the Paris Climate Summit in 2015. It was essentially a trial run of the pledge and review cycles that will occur under Paris. It was also entirely unimpressive. The synthesis report was not a major political or media event and no country has increased their ambition since this initial round of pledges.

For many countries, the pledge and review process has been occurring on a much longer timeframe. For instance, the European Union had a target under the Kyoto Protocol (binding in 2005), submitted a pledge for the Copenhagen conference (2009) and put forward a new pledge prior to Paris (2015). The EU and many developing countries have already been putting forward new climate targets every 5 years. There has also been a de facto review process. Between Kyoto and Paris, the Intergovernmental Panel on Climate Change released reports in 2001, 2007 and 2014. While they do not directly review the adequacy of existing actions, the reports do put forward pathways, some of which include studies on the adequacy of existing trajectories based on policies and pledges.

The pledge and review structure of the Paris Agreement is therefore no novel approach: it is simply a legal ribbon around exiting processes.

There are already numerous non-UNFCCC mechanisms to create pressure on individual states. One well-known initiative is the Climate Action Tracker, which regularly reviews and rates individual country's pledges as "sufficient", "medium" and "insufficient". It also reviews current global efforts and projects these into the future to estimate temperature rise. The Climate Action Tracker is a more accessible and judgemental tool than the global stocktake or transparency framework is likely to be. In addition to these civil society initiatives, there has been a plethora of academic papers that review both country and global climate efforts (Glomsrød, Wei, & Alfsen, 2013; Meinshausen et al., 2015; Riahi et al., 2015; Rogelj et al., 2010).

Despite the abundance of mechanisms for pressure, most governments have not put forward adequate pledges, and many have backtracked on their targets and actions. Canada missed its target under the first commitment period of the Kyoto Protocol. While there was condemnation from peers and civil society, Canada opted to withdraw from the Kyoto Protocol rather than repent and reform.

In 2013, the Australian Government decided to abolish its domestic carbon pricing scheme, leading to a subsequent rise in greenhouse gas emissions. There was international outrage but no change in government position or restatement of the pricing system.

Most recently, the Trump administration has released an executive order that rolls back most of Obama's climate legacy (Davenport, 2017). The USA would have needed full implementation of the Clean Power Plan, as well as additional measures, to meet its already insufficient NDC (Climate Action Tracker, 2017). Trump's actions will most likely result in the USA missing its first pledge and increasing emissions through to 2030 (Höhne et al., 2017; Lux Research, 2016) Once again, this was met with international and domestic expressions of disapproval. Yet the administration has given no indication that they will repeal the executive order, let alone put in place additional and stronger mitigation instruments.

History suggests that the balancing feedback of political pressure portrayed in Fig. 3.2 is weak at best. It has rarely been the case in climate politics, or other international agreements, that the strength of domestic policies has been majorly altered due to political pressure from review processes.

Government actions on climate change have namely been built on a calculus of political and economic costs and benefits. Diplomatic blame and criticism from some domestic constituents will play a part in this calculus. But it is rarely the decisive factor. Major policy changes are unlikely to occur, unless there is the fear of tangible losses. Unless there is a fall in popular support resulting in the loss of power or partners resorting to trade measures. Unfortunately, history suggests that neither other states nor most citizens care deeply enough to mount this sort of pressure on climate issues. If political pressure worked in drastically changing state behaviour, the world would be on a far more favourable emissions trajectory. Paris offers few additional mechanisms to suggest that this feedback mechanism will suddenly become effective.

3.5 Mixed Market Messages

Figure 3.3 is a portrayal of how the investment signal or market message should theoretically work. Investor trust in the long-term goals established under Paris creates a certain perceived level of regulatory certainty. This deters emissions-intensive investments and incentivises low-emissions investments. As the market shifts, decarbonisation appears to be an increasingly credible pathway and trust in long-term goals is strengthened. This establishes a virtuous cycle of heightening investment in low-emissions industries and spiralling confidence in emissions-intensive ones.

This idea of the market message is promising but empirically unconvincing. The world does not have a clear understanding of how international law shapes investment patterns, particularly in the case of legal goals that are not underlined by tangible enforcement mechanisms. There has been some research on how trade and investor-state dispute settlement (ISDS) mechanisms have influenced investment decisions (Schill, 2007; Tienhaara, 2011). Some of the causal links rest on a causal chain whereby international provisions influence national policy which in turn shapes investment patterns and business actions. One example of this is the idea of "regulatory chill" whereby the existence of ISDS clauses cause governments to avoid improving the regulation of public goods due to fears of foreign investor litigation (Tienhaara, 2011). This includes environmental goods and services (Schill, 2007).

Previous findings in trade and commercial law cannot be easily applied to the Paris Agreement. ISDS provisions and outcomes under the World Trade Organization have tangible domestic impacts and are both judiciable and enforceable. The Paris Agreement lacks punitive enforcement mechanisms and is not clearly judiciable.

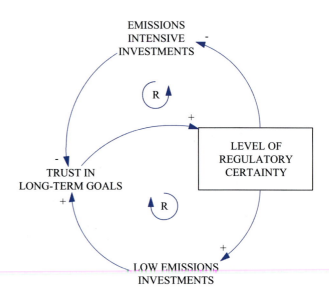

Fig. 3.3 The investment signal feedback

There is no threat of trade sanctions or foreign investor lawsuits to shape government regulatory behaviour and, in turn, investment patterns. The Paris Agreement is qualitatively different and relies on loose long-term goals and nonbinding pledges. Empirically, there are few to no studies on how this type of legal architecture can shape investment patterns. Given this lack of an evidence base, the concept of a market message is a questionable basis for the success of a multilateral agreement. It is even more tenuous given the enormous speed and scale of investment that is needed. To meet the goals of Paris, emissions will need to halve every decade (Rockström et al., 2017).

The history of renewable energy investment suggests that the investment signal from international climate law may not exist at all. Renewable energy alongside coal are the two clearest markets where an investment signal should be visible. Figure 3.4 summarises global renewable energy investment from 2004 to 2016 (Frankfurt School and UNEP, 2016).

Comparing this data to the history of the UNFCCC reveals that there is no clear link between renewable energy investment and outcomes in climate negotiations. In December 2009, the Copenhagen climate conference collapsed, resulting in a vague non-consensual and nonbinding accord that was noted by the parties. Presumably this failure of international climate law would have undermined investor confidence in renewable energy. Instead, in 2010, renewable energy investment increased by 34% to US$239.2 billion. This was an annual increase of US$60.5 billion, the largest increase on record since 2004 (Frankfurt School and UNEP, 2016). The agreement to the Durban Platform of 2011 was widely considered to be a breakthrough that put in place a credible pathway to a new global agreement. Prior to that, the Mexican presidency expertly manoeuvred the world into adopting the Cancun Accords. This was widely perceived to be a significant step forward after the failure

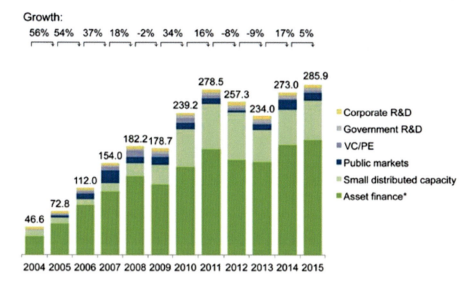

Fig. 3.4 Global renewable energy investment 2004–2015

of Copenhagen. Yet the period of 2011–2013 saw two consecutive years of decreasing investment. These were the highest decreases on record, with –8% in 2011–2012 and –9% in 2012–2013 (Frankfurt School and UNEP, 2016).

The closest sign of an investment signal is the increase in investment after 2005 with the entry into force[2] of the Kyoto Protocol. However, this was part of a more general trend that had begun before 2005. In 2004, when the future of Kyoto was still uncertain, there were record levels of investment growth (56%, or US$26.2 billion). At least in the short term, renewable energy investment was not determined, or seemingly significantly influenced, by international legal outcomes.

Recent post-Paris history further undermines the idea of the investment signal. New investment in clean energy was down by 18% in 2016. It has fallen a further 17% in the first quarter of 2017. Part of this is due to the plummeting capital cost of renewable energy, meaning less finance is required for installations. The larger reason is the rolling back of government support for solar and wind projects in both China and the USA (Shankleman, 2017).

The Paris Agreement did not turbocharge the renewable energy industry by providing regulatory certainty. Instead, market trends were shaped by domestic policy choices and wider technological trends. Similarly, the significant growth in renewable energy from 2004 to 2008 was driven by the actions of the EU, in particular the feed-in tariff of Germany. European actions at this stage were partly influenced by the adoption of the Kyoto Protocol. But the predominant driver was hard-nosed

---

[2] "Entry into force" refers to the agreement becoming legally binding and operational. In the case of the Kyoto Protocol, this required at least 55 ratifying countries accounting for 55% or more of regulated greenhouse gases.

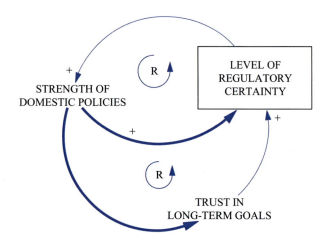

Fig. 3.5 The investment signal in practice

national interest. For instance, support for renewable energy in Germany has been motivated by the desire to develop new sources of economic comparative advantage rather than global altruism. Concrete regulatory changes from the national government have shaped trends in renewable energy investment, not international law or global goals.

Figure 3.5 provides a more realistic depiction of how investment patterns are shaped by international law. The strength of domestic policies is the key determinant of both the trust placed in long-term goals and the level of regulatory certainty. Hard-nosed, self-interested investors are likely to place far more emphasis on the short-term actions and pledges of countries than ambiguous international goals set for the latter half of the century. This means that if domestic policies are consistent and ambitious there will be regulatory certainty that feeds back into greater action. But weak policies and/or inconsistent action will undermine certainty in both regulation and the achievement of long-term goals.

Evidence to date suggests that investors are placing much more faith in the existing actions of government than the goals of the Paris Agreement. The investment signal, if it does exist, is clearly insufficient to cause a market shift at the scale and speed needed.

3.6 Locked-in

A key dimension of any system is not just the feedbacks involved, but the rate of change.

The Paris Agreement is too little, too late. Recent studies have found that the timing of emissions reductions is the most important factor when it comes to the success of global mitigation. In modelling, an early start to mitigation is more significant to long-term success than the use of negative emissions technologies or the price on carbon (Hatfield-Dodds, 2013; Rogelj, 2013). A delay and continuation of existing efforts up until 2025 almost rules out the 2 °C target (Rogelj, 2013). Under Paris, the first global stocktake is in 2023, and national pledges do not need to be updated until 2025. By the time the pledge and review system is intended to take effect, the world will likely already be locked into an emissions trajectory of more than 2 °C.

Delay is fatal largely due to the "lock-in" of carbon-intensive infrastructure. One recent study from Oxford University calculated that no new fossil fuel electricity generation could be built after 2017 (Pfeiffer, Millar, Hepburn, & Beinhocker, 2016). All additional fossil fuel-based electricity generators built after this would need to be prematurely retired to meet the 2 °C target. As of 2016, 58 Gt CO_2 worth of new coal-powered generators were in construction, and a further 186 Gt CO_2 had been announced and permitted (Shearer, Ghio, Myllyvirta, Yu, & Nace, 2017). This pipeline is made up of 3697 generating units globally, worth US$981 billion. These plans will have an average lifespan of 40 years (Shearer et al., 2017). To be compatible with Paris, almost all will need to be prematurely retired or stranded at a much earlier date.

The same lock-in effect applies to fossil fuel. One recent study by McGlade and Elkins found that in order to stay within 2 °C, more than 80% of coal reserves, half of gas reserves and a third of oil reserves need to stay underground (McGlade & Ekins, 2015). It should be noted that "reserves" refers to existing resources that are known and currently economically recoverable. This makes any further exploration for fossil fuels incompatible with the goals of the Paris Agreement (Carbon Tracker Initiative, 2013).

Yet the Agreement has no way of officially deterring or influencing such investments. There are no provisions to address fossil fuel supply. The entire treaty does not mention coal, oil or fossil fuel once. Since the adoption of the Agreement, Canada has announced the construction of two new major oil pipelines, the USA has moved to streamline the approval process for the Keystone XL oil pipeline and the Australian Government is looking to subsidise a major new coal development—the Adani coal mine—with A$1 billion.

The Paris Agreement has done nothing to deter the lock-in of a new fossil fuel supply. If there is an investment signal, then clearly governments and the fossil fuel industry have not heard it.

The goals of Paris can still be met, even with the existing timeframes and delay of effective action. However, if they are to be met, there will likely need to be a significant amount of stranded assets. One modelling exercise conducted prior to the Paris Agreement by the Potsdam Institute noted that "huge quantities of installed coal capacity will need to be prematurely retired between 2030 and 2050. Such a vast global write-off of capital would be unprecedented in scale" (Bertram et al., 2015). The Paris Agreement has seemingly done little to change this.

Delay makes emissions reductions in the future costlier, creates stranded assets, restricts technological options and passes on much of the effort to future generations (Kemp & Jotzo, 2015). Currently, reducing emissions is cheap and can even result in a net benefit (Jotzo & Kemp, 2015). Yet governments have still proved unwilling to take adequate action. It is difficult to see mitigation heightening to the necessary level with higher costs and stranded assets.

The Paris Agreement contained the seeds of its own demise. Due to the timeframe embedded into its implementation, significant delay is already committed into the system.

3.7 Conclusions

The pledge and review system of Paris is neither timed nor structured to meet its own targets. There are potential feedback mechanisms through increasing political pressure and creating regulatory certainty for low-carbon investments. However, neither of these are empirically persuasive or historical supported. Even if the feedbacks did exist and were strong enough to sufficiently increase global action, both lack the right provisions in the Agreement to make them effective.

The heady optimism over political pressure and the "investment signal" seem more to be ex post rationalisations of Paris rather than premeditated design features to make Paris successful over time. They are reasoned justifications of an agreement that prioritises breadth over depth and participation over structure. These architectural choices were not created to trigger certain feedbacks, but to appease certain actors.

Paris was designed largely for political purposes that suited the interests of the USA. The overriding priority of Paris was never dynamic effectiveness, but universal participation. The mandate of the Durban Platform for Enhanced Action was to "develop a protocol, another legal instrument or an agreed outcome with legal force under the Convention applicable to all Parties". The emphasis on universal participation was necessary for both a consensus-based process and to appeal to the USA. Since the Byrd-Hagel resolution of 1997, the USA has repeatedly made it clear that its participation rests on the involvement of major emerging emitters. Universality was a precondition for US buy-in. The influence of the USA is apparent in more direct ways as well. The use of nonbinding pledges and the absence of legal obligations on financing were all implemented so that the USA could legally participate without the consent of the US Senate (Kemp, 2016). The Paris Agreement was shaped so that the USA could be involved purely through the approval of President Obama. Yet the USA has already indicated that it will withdraw from the Agreement under President Trump.

Ultimately, the system of the Paris Agreement has been shaped to suit the interests of the USA, not to create a dynamic and effective international architecture. A systems critique of the Paris Agreement reveals that while there are feedback mechanisms they are unlikely to work. Neither political pressure nor investment signals are proven leverage points for change. Moreover, the Agreement has not been designed to adequately drive either of these feedbacks.

This is not to say that the Paris Agreement cannot or will not succeed. The current pace of low-carbon technological change could overwhelm lock-in dynamics and political inertia. The development of saviour technologies such as negative emissions technologies could also bring the world back from the brink. But these are technological waves that were occurring prior to Paris. The Paris Agreement will not be the decisive factor in whether a clean energy revolution or negative emissions deus ex machina occurs. The universal nature of Paris gives it symbolic strength. But its structural design to achieve universality makes it weak as a system to increase climate action. As an agreement and a system, it is universal, but largely useless.

References

Aldy, J. E., Barrett, S., & Stavins, R. N. (2003). Thirteen plus one: A comparison of global climate policy architectures. *Climate Policy, 3*(4), 373–397.

Australian Government. (2015). *Australia's intended nationally determined contribution to a new Climate Change Agreement*, Canberra, Australia. Retrieved from http://www4.unfccc.int/submissions/INDC/PublishedDocuments/Australia/1/AustraliasIntendedNationallyDeterminedContributiontoanewClimateChangeAgreement-August2015.pdf

Bertram, C., et al. (2015). Carbon lock-in through capital stock inertia associated with weak near-term climate policies. *Technological Forecasting and Social Change, 90*(PA), 62–72. https://doi.org/10.1016/j.techfore.2013.10.001.

Bodansky, D. (2015). Legally binding versus non-legally binding instruments. In S. Barrett, C. Carraro, & J. de Melo (Eds.), *Towards a workable and effective climate regime* (pp. 155–167). London, UK: CEPR Press. Retrieved from http://voxeu.org/sites/default/files/image/FromMay2014/Climatechangebookforweb.pdf.

Bodansky, D. (2016). The Paris Climate Change Agreement: A new hope? *The American Journal of International Law, 110*(2), 288–319.

Carbon Tracker Initiative. (2013). *Unburnable carbon 2013: Wasted capital and stranded assets* (pp. 1–40). London: Carbon Tracker & The Grantham Research Institute, LSE. Retrieved from http://www.carbontracker.org/site/wastedcapital.

Climate Action Tracker. (2017). *USA. Climate Action Tracker*. Retrieved March 27, 2017 from http://climateactiontracker.org/countries/developed/usa.html

Davenport, C. (2017). Trump signs executive order unwinding Obama climate policies. *New York Times*. Retrieved July 12, 2017 from https://www.nytimes.com/2017/03/28/climate/trump-executive-order-climate-change.html

Frankfurt School and UNEP. (2016). *Global trends in renewable energy*, Frankfurt, Germany. Retrieved from http://fs-unep-centre.org/sites/default/files/publications/globaltrendsinrenewableenergyinvestment2016lowres_0.pdf

Glomsrød, S., Wei, T., & Alfsen, K. H. (2013). *Pledges for climate mitigation: The effects of the Copenhagen accord on CO_2 emissions and mitigation costs* (pp. 619–636). New York: Springer.

Harvey, F. (2015). Paris climate change agreement: the world's greatest diplomatic success. *The Guardian*. Retrieved July 12, 2017 from https://www.theguardian.com/environment/2015/dec/13/paris-climate-deal-cop-diplomacy-developing-united-nations

Hatfield-Dodds, S. (2013). Climate change: All in the timing. *Nature, 493*, 35–36. Retrieved from http://www.nature.com.proxy.library.vanderbilt.edu/nature/journal/v493/n7430/full/493035a.html%5Cnhttp://www.nature.com.proxy.library.vanderbilt.edu/nature/journal/v493/n7430/pdf/493035a.pdf

Heywood, P. (2013). Climate change negotiations—legally binding treaty or voluntary agreements: Lessons from international action on tuberculosis. *Wiley Interdisciplinary Reviews: Climate Change, 4*(6), 467–477. https://doi.org/10.1002/wcc.237.

Höhne, N. et al., 2017. "Action by China and India Slows Emissions Growth, President Trump's Policies Likely to Cause US Emissions to Flatten", Climate Action Tracker.

Jacquet, J., & Jamieson, D. (2016). Action by China and India slows emissions growth, President Trump's policies likely to cause US emissions to flatten, Soft but significant power in the Paris agreement. *Nature Climate Change, 6*(7), 643–.646. Retrieved from http://www.nature.com/doifinder/10.1038/nclimate3006

Jotzo, F., & Kemp, L. (2015). *Australia can cut emissions deeply and the cost is low.* Canberra, Australia: Australian National University.

Kemp, L. (2016). Bypassing the "ratification straitjacket": Reviewing US legal participation in a climate agreement. *Climate Policy, 16*(1), 1011–.1028. Retrieved from http://www.tandfonline.com/doi/full/10.1080/14693062.2015.1061472

Kemp, L., & Jotzo, F. (2015). *Delaying climate action would be costly for Australia and the world.* Canberra, Australia: Australian National University.

Lux Research. (2016). *A Trump presidency could mean 3.4 billion tons more U.S. carbon emissions than a Clinton one.* Retrieved March 1, 2017 from http://www.luxresearchinc.com/news-and-events/press-releases/read/trump-presidency-could-mean-34-billion-tons-more-us-carbon

McGlade, C., & Ekins, P. (2015). The geographical distribution of fossil fuels unused when limiting global warming to 2 °C. *Nature, 517*(7533), 187–.190. Retrieved from http://www.nature.com/doifinder/10.1038/nature14016%5Cnpapers2://publication/doi/10.1038/nature14016

Meinshausen, M., Jeffery, L., Guetschow J, du Pont Y R., Rogelj J, Schaeffer M., ... Meinshausen, N. (2015). National post-2020 greenhouse gas targets and diversity-aware leadership. *Nature Climate Change, 1500*(October), 1–10. Retrieved from http://www.nature.com/doifinder/10.1038/nclimate2826

Pfeiffer, A., Millar, R., Hepburn, C., & Beinhocker, E. (2016). The 2C capital stock for electricity generation: Committed cumulative carbon emissions from the electricity generation sector and the transition to a green economy. *Applied Energy, 179*, 1395–1408. https://doi.org/10.1016/j.apenergy.2016.02.093.

Rajamani, L. (2016). Ambition and differentiation in the 2015 Paris Agreement: Interpretative possibilities and underlying politics. *The International and Comparative Law Quarterly, 65*(2), 493–514. Retrieved from http://www.journals.cambridge.org/abstract_S0020589316000130

Raustiala, K. (2005). Form and substance in international agreements. *The American Journal of International Law, 99*(3), 581–614.

Riahi, K., et al. (2015). Locked into Copenhagen pledges—Implications of short-term emission targets for the cost and feasibility of long-term climate goals. *Technological Forecasting and Social Change, 90*(PA), 8–23.

Rockström, J., Gaffney, O., Rogelj, J., Meinshausen, M., Nakicenovic, N., & Schellnhuber, H. J. (2017). A roadmap for rapid decarbonization. *Science, 355*(6331), 1269–1271. https://doi.org/10.1126/science.aah3443.

Rogelj, J. (2013). Earth science: A holistic approach to climate targets. *Nature, 499*(7457), 160–161. https://doi.org/10.1038/nature12406.

Rogelj, J., den Elzen, M., Höhne, N., Fransen, T., Fekete, H., Winkler, H., et al. (2016). Paris Agreement climate proposals need a boost to keep warming well below 2 °C. *Nature Climate Change, 534*(7609), 631–639. https://doi.org/10.1038/nature18307.

Rogelj, J., Nabel, J., Chen, C., Hare, W., Markmann, K., Meinshausen, M., et al. (2010). Copenhagen Accord pledges are paltry. *Nature, 464*(7292), 1126–1128. https://doi.org/10.1038/4641126a.

Schill, S. W. (2007). Do investment treaties chill unilateral state regulation to mitigate climate change? *Journal of International Arbitration, 24*(5), 8. Retrieved from http://0-www.kluwerarbitration.com.alpha2.latrobe.edu.au/print.aspx?ids=ipn28069

Shankleman, J. (2017). Clean energy investment drops 17% as China and U.S. Scale Back. *Bloomberg.* Retrieved April 24, 2017 from https://www.bloomberg.com/news/articles/2017-04-11/clean-energy-investment-drops-17-as-china-and-u-s-scale-back

Shearer, C., Ghio, N., Myllyvirta, L., Yu, A., & Nace, T. (2017). *Boom and bust 2016: Tracking the global coal plant pipeline.* Washington DC: Greenpeace.

The Economist. (2015). The Paris agreement marks an unprecedented political recognition of the risks of climate change. *The Economist.* Retrieved from http://www.economist.com/news/international/21683990-paris-agreement-climate-change-talks

Tienhaara, K. (2011). Regulatory chill and the threat of arbitration: A view from political science. In C Brown, K Miles (Eds.), *Evolution in Investment Treaty Law and Arbitration* (pp. 606–627). Cambridge: Cambridge University Press. Retrieved from http://ssrn.com/abstract=2065706

UNFCCC. (2015). *Paris Agreement*. Paris, France: United Nations.

UNFCCC. (2017). *United Nations framework convention on climate change: The Paris Agreement*. Retrieved from http://unfccc.int/paris_agreement/items/9485.php

van Asselt, H. (2016). International climate change law in a bottom-up world. *Questions of International Law, 26*, 5–15.

Chapter 4
Structural Impediments to Sustainable Development in Australia and Its Asia-Pacific Region

Ahmed Badreldin

Abstract In its efforts to administer a collective global response towards combating climate change and limiting global warming, the United Nations, at its 2015 Climate Change Conference, succeeded in committing member nations to the Paris Agreement. Through the Agreement, the United Nations exemplified its dedication to supporting sustainable development. Accordingly, the primary objective of this chapter is to investigate structural impediments that prevent Australia and the Asia-Pacific region from achieving their Paris Agreement targets and consolidating sustainable development. While neoliberal globalisation has nurtured ecological damage and widespread poverty and wealth inequality in a systematic manner, this chapter argues that the accumulation and persistence of these structural deficiencies portend severe implications against the attainment of sustainability targets. The chapter introduces an assessment approach suggesting the stage of economic development, the extent of social equity and the political orientation of each country to distinguish its vulnerability and exposure to these structural impediments. It further addresses difficulties that governments, businesses and civil societies face with a focus on solving them. Lastly, it anticipates a paradigm shift away from the GDP growth-based, fossil fuel-driven industrial type of economic development towards a more inclusive and equitable model comprising eco-efficient low-carbon enterprises and economies. The chapter concludes that only the equitable, more inclusive and democratic developmental regimes are capable of consolidating sustainable development.

Keywords Sustainable development • Structural impediments • Corporatocracy • Democracy • Australia

A. Badreldin (✉)
University of Newcastle, Callaghan, NSW, Australia
e-mail: a.badr41@yahoo.com

4.1 Introduction

The rapid diffusion of the economic slowdown accompanying the 2008 global financial crisis affirms the transnational phenomenon of modern globalisation. The broad scope and fast spread distinguishing this crisis strongly reflect an unprecedented extent of global interdependence and interconnectivity (McGrew, 2013; Mészáros, 2012). Increasingly perceived as a systematic crisis of global capitalism, the global financial crisis further affirms inherent failures in the neoliberal capitalism detrimental to the attainment of sustainable economic development (Amin, 2012; Mészáros, 2012). Indeed, the intensifying ecological damage; energy, food and water shortages; widespread poverty and unemployment; declining economic growth rates; uneven geographical economic development; accelerating wealth gaps; and cyclical financial crises crystallise persistent structural market failures.[1] These failures are commonly associated with the contemporary paradigm of corporate-driven global capitalism, or "corporatocracy" (Amin, 2012; Anderson & Bows, 2012; Dafermos, Galanis, & Nikolaidi, 2015; Stiglitz, 2006; Harvey, 2011) Today, such systematic imbalances and dysfunctionality have accumulated and deepened core impediments that act against the attainment of a more sustainable form of global economic development, whether in Australia or elsewhere (Kotz, 2009; Speth, 2008).

Fortunately, there is another side of the story. Contemporary globalisation has also brought about notable progress in almost every aspect of human life. The elevation of innovation and entrepreneurship, particularly in information and communication technologies, financial markets and transportation, has unleashed unparalleled expansion in the medium of global interactions and trade, dissolved geographical borders and diluted cultural authenticities beyond previous magnitudes (Amos, Keiner, Proske, & Radtke, 2002; Stiglitz, 2006). The easier access to global resources and markets has enabled many successful businesses emerging from local domains to extend their productive activities, investments, innovations and expertise, thereby enhancing global welfare and interdependence (Schäuble, 2017). This contemporary global interdependence prompts dialogue around joint actions when confronting mutual challenges. Nevertheless, determined efforts by the United Nations (UN) to address such challenges in its 2015 Climate Change Conference in Paris have targeted committing its member nations to combat climate change and limit average global temperature increases. Within the UN's wider context of its sustainable development agenda, this chapter anticipates structural challenges to sustainability in Australia and the Asia-Pacific region.

---

[1] Structural/systematic impediments, failures, challenges or deficiencies are terms used interchangeably within the chapter to identify core and fundamental dysfunctionalities that are not resolvable by minor modifications such as, for example, state bailouts or new regulations. However, only major structural interventions are applicable, which implies a replacement rather than series of corrections or adjustments to the overall design and framework; see Kotz (2009) and Mészáros (2012).

Comprehensive analysis of structural impediments advocates both a global approach and a nation-based focus. The contemporary global interconnectivity and interdependence induce extending the domain of analysis to incorporate both foci. This chapter proposes that each country's unique stage of economic development, the extent of social equity and the political orientation will determine the degree of exposure and vulnerability to these impediments. Acknowledging that these structural deficiencies are part of corporate-driven neoliberal globalisation necessitates exploring the contradictions between the rhetoric of contemporary neoliberalism and its realities. Addressing the discrepancy between the present regime of neoliberal globalisation and the consolidation of sustainable development dictates tackling several issues involving sustainability challenges against economic development, democracy and cultural conflicts associated with inducing lifestyle and behaviour changes. It further includes the role of governments, businesses and community-driven nongovernment organisations (NGOs) in reducing ecological footprints and restoring environmental damage. Finally, the chapter anticipates the current heightening of economic, social and political instabilities to invoke an inevitable paradigm shift away from the GDP growth-based/fossil fuel-based developmental model towards a more sustainable, socially just and eco-efficient global economy (Anderson & Bows, 2012; Elkington & Hartigan, 2013; McIntosh, 2009).

4.2 Corporatocracy Between Theory and Practice

Corporatocracy, resembling the contemporary paradigm of corporate-driven neoliberal globalisation, appears clearly fraught with structural contradictions (Bivens, 2011; Mészáros, 2012). A way of analysing such contradictions is to seek theoretical foundations and practical realities of the neoliberal component that gave rise to those inconsistencies. Following the objectives of this study, this chapter applies Michel Foucault's analytics of governmentality,[2] as introduced by Mitchell Dean in his book Governmentality: Power and Rule in Modern Society, to assess the evolution of the global neoliberal regime of governance and capture its fundamental contradictions. With the intention of dismantling the liberal and neoliberal theories and concepts, assessing their advent following the welfare state paradigm marks a logical starting point.

---

[2] Analytics of governmentality refers to analytical tools applied in studying the rationalities, regularities, logic, strategies, mechanisms, as well as economic orientations that shape the identity of regimes of governance. More importantly, it extends beyond claimed, presumed or declared values of governance and enables detecting embedded systematic impediments and challenges. It facilitates exposing and solving what Max Weber describes as "inconvenient facts"; see Dean (2009, p. 48). These are areas in systems that appear to malfunction or provoke resistance to change and reformation. By becoming familiar with these complexities, Foucault suggests analytics of governmentality contributes towards highlighting and removing concealed problematic causes and reveal possible formulation of alternatives that otherwise might have been undetectable or "taken for granted"; see also Dean (2009), p. 50).

Building on Foucault's concepts of governmentality, historical reviews reveal that the oppressive uniformity inherent in bureaucratic over-governance of the 1960s and 70s, along with the hierarchical and paternalistic approach to social control, has rendered the welfare state model irresponsive to the varying needs of both individuals and communities. The accumulated public and political rejection of this uniformity apparently advocated the replacement of the welfare rationality with more liberal forms of governance. This refusal paved the way for a transformation of governmentality, which Foucault asserts to have created a quasi-natural reality intended to circumvent the excessive dependency, inefficiency, rigidity and unaccountability characterising the welfare system. Its replacement, the liberal art of governance, assumed more efficient resource management by limiting excessive state influence and intervention in favour of professed individual rights and liberties. The aim is to transform institutional and cultural rationalities and practices towards reconfiguring the concept of freedom of subjects as free consumers and enterprises in free markets. Empowered by their skills and expertise, individuals are to serve as labour in those same markets. Subjects acting individually or collectively are to bear multiple responsibilities for themselves against the risks and life challenges of, for example, illness, poverty, unemployment and illiteracy.

Apart from the appealing explicit rhetoric in the founding principles of the contemporary global neoliberal paradigm, critical contradictions have arisen accompanying the transformation of these theories and concepts into practice. With time, liberalism has developed visible authoritarian features and, in some circumstances, modern liberal democracy has slid into the paternalistic conduct and reinscribed coercive elements of fascism. Theoretically, liberal democracies advocate plurality while operating through the freedoms of the ruled. This notion as it stands suggests the ruled have a capacity to act otherwise when conducting themselves or their activities. However, in practice, the capitalist labour markets of classical liberal governance have made working men reliant on those markets for work and pay, and reflexively, women and children of these men dependent on those same markets. Contrary to the seductive liberal rhetoric of greater rights, liberties and freedom of choice for all, along with promises of freely functioning and wealth-generating trade and markets, the reality has conveyed uneven, declining and redistributive growth yielding a concentration of wealth into fewer hands (Gills & Rocamora, 1992; Hedges, 2010).

4.3 Structural Socioeconomic Impediments

Upon the seizure of the high growth rates sustained in the aftermath of the Second World War, and further exacerbated by the mid-1970s oil shock recession, the models of the welfare state succumbed to neo-Austrian politicoeconomic configurations. Previously held back by communist–capitalist struggles, the rise of the industrial and technological powers of many Western economies vis-à-vis their communist counterparts paved the way for the stronger free market economies to abandon the Keynesian component that had for decades moderated its radical laissez-faire market inclinations (McIntosh, 2009; Robinson, 2004). Efforts to attract investments and achieve higher GDP have tempted many governments to ignore the quality and distribution of the economic proceeds. This paradigm of corporate-driven global capitalism has tailored, sponsored and, in many instances, enforced capital-friendly economic agendas permitting relaxed access to local resources and markets (Amin, 2012; Harvey, 2007). The pursuit of profitability as the single bottom line while being unaccountable to the societies in which they operate prompts transnational corporations to disregard and circumvent the externalities associated with their economic activities. It comes then as no surprise that the general public who least benefit from such activities ordinarily endure those externality costs (Madeley, 2003; Minio-Paluello, 2014).

Constrained by the mid-1970s global oil shock recession and prompted by schemes of market liberalisation and structural adjustment programmes, the world was set on steadily embracing an aggressive capitalist agenda. This agenda has rendered a considerable shift in the balance of power away from national states in favour of private businesses, financial institutions and transnational corporations (Castells, 2011). In fact, what emerged to become today's neoliberal order is mere protraction of the Washington Consensus and its International Monetary Fund's standard policy prescriptions. These prescriptions, or rather instructions, involve financial deregulations, currency devaluations and policy-based lending in conjunction with austerity measures comprising systematic cuts in state budgets and underfunding of social safety nets. Commonly, this is a part of wider policy agenda that comprises loosening labour protection legislation and price control mechanisms, commodifying natural resources and privatising state-owned enterprises (Gills & Rocamora, 1992; Harvey, 2007). The shrinking capacity of the states to deliver education, healthcare, employment, housing and transportation services has afforded a fertile soil for the private sector to undertake responsibilities previously held by those states. The main concern is that the disposition of public assets has reflexively deprived national states of substantial capability to influence market outcomes, particularly for necessity goods and services that entail a social dimension, in favour of an absent security holder that the states have no control over (Amos et al., 2002; Anderson & Cavanagh, 2004).

Typical neoliberal policy prescriptions involve privatisation of public assets, which may otherwise compete against private capital or limit their monopoly powers. Similarly, the relaxation of price control mechanisms, the ease of health and safety regulations and the reduction and redirection of public social welfare, subsidy provisions and tax deductions to compensate for corporate tax breaks and bailouts rationally favour profit-seeking business dealings. The adoption of export-led growth over import substitution undermines national protectionism and, accordingly, eases the disposition of natural capital and the penetration of local markets. The liberalisation of financial markets regularly follows to ensure profit repatriations, whereby the deregulation of capital movement permits transnational profit-seekers to practise predatory and speculative financial dealings. The list of neoliberal policy directions further extends to include the replacement of communal property rights with corporate protectionism and intellectual property rights and patents that rationally protect the competitiveness of innovative multinational enterprises, while disregarding the social costs and benefits involved (Stiglitz, 2006). Policies aimed at undermining the power relations of labour unions and disciplining domestic labour entail two directions. The first is the adoption of the natural rate of unemployment to replace full employment thus deliberately generating a pool of labour surplus or a "reserve army of unemployed" as identified by Marx (Watts & Mitchell, 2012, p. 229). The second involves loosening immigration policies to attract skilled foreign labour or outsourcing economic activities to countries with relatively low-cost labour markets. Indeed, these typical neoliberal policy interventions systematically expose ordinary citizens to global market mechanisms and make them dependent on profitseekers to meet their basic needs (Gills & Rocamora, 1992; Harvey, 2007; Kotz, 2009; Madeley, 2003).

Failure to secure inclusive trickle-down economic proceeds, along with the growing reliance on market-oriented dynamics in determining the price and availability of such goods, has led to a mounting exclusion of some social segments unable to afford market rates. Furthermore, this assumed corporatocracy has contributed to more assets in fewer hands. Those favoured few have gradually exploited monopolistic powers over local markets rather than enhancing market competitiveness and efficiency or maximising holistic welfare as initially sought from such policy directions (Reich, 2016). Apparently, such domination over the socioeconomic realm has proved dysfunctional in a systematic manner and, consequently, detrimental to sustainable development (Kothari, Demaria, & Acosta, 2015; Kotz, 2009). In conclusion, only the equitable and more inclusive socioeconomic developmental models are capable of consolidating sustainable development.

4.4 Structural Political Impediments

It was around the early 1980s when the neoliberal policy regimes came to dominate the highest political ranks as crystallised by the inauguration of Thatcher in the UK and Reagan in the USA, thereby declaring a new era of global governance and capital power (Davies, 2004; Harvey, 2007). Over an extended period of almost four decades of corporate-driven neoliberal globalisation, the swelling powers of profit-seeking lobbyists persistently challenged the sovereignty of representative governments and their environmental welfare (Anderson & Cavanagh, 2004; Friedman, 2013). Empirical research justifies such disposition by affirming corporatocracy to exhibit wide wealth gaps and income inequalities due to its inclinations of reducing social spending and public services (Ayeb & Bush, 2014; Harvey, 2007; Piketty, 2014). In light of this mounting global inequality, Alesina and Perotti (1996) assert that inequality is inversely related to investment inflows necessary for assuring sociopolitical stability. And as investments comprise a major driver for growth, corporatocracy appears to subvert its rhetoric of achieving durable growth and development. Indeed, the consolidation of democracy is more probable where the degree of inequality is narrowed. However, high inequality is destabilising, particularly for its tendency to trigger social tensions and disintegration (Acemoglu & Robinson, 2001; Kristof, 2014; Stiglitz, 2013; Wayne Nafziger & Auvinen, 2002). Nevertheless, in a capitalist global domain where differential economic and financial capacities reflect equivalent political power relations, concepts of capitalism and democracy are deemed highly incompatible (Esposito, 2014; Piketty, 2014; Swift, 2002).

In a fully functioning (representative) democratic system, where the median voter sets tax rates, public spending, as well as the distribution of economic rent, burdens and developmental proceeds, the median voter is assumed to bring about policy directions but is reluctant to any capital domination (Posner, 1997). Contrarily, with the contemporary less-restrained corporatocracy, modern formal democracies are refashioned into what Gills and Rocamora (1992, p. 501) identify as a "low-intensity" procedural process comprising nominal political contestations within a controlled, procedural, semi-democratic political order (Acemoglu & Robinson, 2001; Amin, 2012; Brysk, 2002; Robinson, 2004). Commonly, such a defective political construct is employed to whitewash flawed regimes and legitimatise the imposition of austerity measures, heavy debt servicing and privatisation schemes, thereby precluding radical politicoeconomic directions or wide-scale public revolt. However, the incapacity of procedural/electoral forms of democracy to render sound legitimacy and sustain stability resides in the fact that these types lack core democratic values of equal participation and representation based on plurality, inclusiveness, mutual accountability and the public's right of self-determination. Ironically, in the name of economic efficiency, this corporatocratic ideological agenda provokes economic disorder and social fragmentation and hence sets the stage for subsequent "ungovernability"(Gills & Rocamora, 1992 p. 507; Robinson, 2004).

Marked by the end of the oil boom and parallel to the explosion of third world debt crisis of the mid-1980s, swelling internal and external debts have progressively imposed huge liabilities on national states and challenged their autonomy and sovereignty by restraining powers of representative governments. The typical scenario contends that governments that hold heavy public debts would necessarily concede to capital-friendly economic agendas dictated by large creditors such as the International Monetary Fund, the World Bank or other financial institutions and their credit rating agencies (Corroto, 2012; Stiglitz, 2006; Swift, 2002). Neither elected by nor accountable to the public, those large creditors further drag the global political order towards their undemocratic orientation (Harvey, 2007). Richard Swift (2002) justifies the inclination of large creditors to oppose sustaining economic and financial policies, particularly public spending on healthcare and education, as they take priority over debt repayments regardless of the long-term economic, social or environmental repercussions.

As an extension of the same argument, securing political financing and heavily funded political campaigns have become a fundamental precondition for electoral success. Such modern-day political order reserves the centres of public decision-making mainly for profit-oriented business elites comprising corporate, financial institutions' and media channels' representatives and owners, all of whom would act in their best self-interests (Friedman, 2013; Herman & Chomsky, 1988; Swift,

2002). Similarly, financing political campaigns legitimatises foreign capital to vote with its dollar bills and lobby against representative governments. Another mechanism involves the displacement of foreign direct investment towards lesser-taxed or regulated markets, thereby forcing democratic governments, if they exist, to compete for investments and a share of exports in a race to the bottom (Amin, 2015).

Here, several critical structural impediments against functioning democracy clearly emerge and persist. Firstly, contemporary formal democracies nurture pro-capitalist politicians capable of successfully raising such financing, yet deny the majority access to influential political posts (Wolff, 2012). Reflexively, such an arrangement disconnects politicians from public welfare and drives the public to either withdraw from the political scene or, worse, vote against the democratic constituent itself. Secondly, business elites and lobbyists are progressively replacing political parties while shaping public debates and policies, which undermines a fundamental pillar of a democratic system of plurality and equal opportunity of political participation and representation. Thirdly, a common challenge to current forms of liberal democracy arises when governments jeopardise long-term sustainability objectives by pursuing a short-sighted opportunistic approach of exploiting immediate benefits to enhance their chances of electoral success (Cadman et al., 2015). However, this study suggests two solutions that appear most relevant in the attempt to overcome such a short-term political cycle dilemma. One is to constitutionalise sustainable development goals. Otherwise, the public may enforce an electoral pledge on political parties to abide by strict sustainability targets as a precondition for political contestation.

Several attempts have been conducted aiming to preserve functioning democracy from capital domination. For example, Samir Amin (2012) emphasises the liberation from capital to be a precondition for progress. A viable mechanism is to set limits on political financing to avoid manipulating the political construct. Alternatively, in his book The Revolt of the Elites, Christopher (1995) conditions the construction of a democratic society that is invulnerable to the distorting influence of economic powers on restricting the unlimited accumulation of wealth itself. In conclusion, regardless of which ideal mechanism shall balance the relationship between capital and politics, only representative democratic forms of governance are capable of consolidating sustainable development.

4.5 Structural Impediments to Sustainable Development in Australia and the Asia-Pacific Region

The Asia-Pacific is a widely diverse region based on the stage of economic development, social equity and political orientation of its separate member nations. This chapter introduces an assessment methodology that acknowledges structural impediments associated with contemporary neoliberal globalisation to restrain each country from achieving its sustainability targets. However, the stage of economic development, wealth equality and political orientation of each country shall distinguish its vulnerability and exposure to these impediments. It follows, therefore, that the more economically developed, socially just and democratic a country is, the less susceptible to structural impediments this country becomes. In the Asia-Pacific region, the study maps some countries against the economic, social and political variables as introduced by the methodology. For example, Japan, South Korea, Singapore and Taiwan stimulated periodic economic growths despite medium inequality (except Singapore, where it is high) and flawed democracy. However, China secured significant economic growth despite relatively high wealth inequality and authoritarian governance. India achieved significant economic growth, despite medium wealth inequality and flawed democracy. In the best position, Australia and New Zealand sustained economic development with stable democracies despite medium wealth inequality (Harvey, 2007; The Economist Intelligence Unit, 2017).

Switching the focus to Australia, according to its 2055 intergenerational report, Australia faces some critical structural impediments while addressing sustainability, the most prominent being climate change (Australian Treasury, 2015). Indeed, Australia's per capita carbon footprint is unsustainable, suggesting its structural transformation shall encounter considerable hardship, particularly with ecological sustainability (McIntosh, 2009; Smith, 2007). The catastrophic impact of climate change to cause an acidic ocean, coral bleaching, rising sea levels and coastal erosion imposes high risks of irreversible damage to its marine life. While around 80% of Australians reside around coastal areas, rising sea levels do, in fact, expose millions of Australians to a significant risk of displacement (McInnes et al., 2016). Moreover, its poor soil (only 6% suitable for agriculture) along with the scarcity of accessible water sources (Commonwealth of Australia, 2014) dictates immediate structural transformation that would ensure preserving or, better, restoring its deteriorating natural capital.

In an attempt to anticipate structural challenges threatening Australia from achieving its Paris Agreement targets and attain sustainable development, this chapter proposes that cultural conflicts of inducing lifestyle and behaviour change constitute the most structured challenge. Individualistic morality has replaced the sense of community, mateship, solidarity, belonging and collective responsibility (Corroto, 2012). The emergence of such individualism, although disrupting to self-sustaining local communities, is alien to the capitalist interests of fostering conspicuous consumption so as to set the mode of production to economic growth rather than the logical opposite as denoted by Marx (Amin, 2012; Kaboub, 2011). Advertising and corporate branding propaganda have promoted consumption, symbolising a psychological quest for affluence, distinctive lifestyle and social advancement, thereby justifying unlimited wealth accumulation (Clifton & Maughan, 2000; Kristof, 2014; Speth, 2008). The detrimental economic costs of this consumerist ideology at the individual level emerge in undermining the intrinsic values of hard work and, in turn, induce a sense of incapability, insecurity, anger, bigotry, alienation and extremism (Kristof, 2014; Madeley, 2003). Implications at the community level comprise systematic destruction of public ownership and depletion of non-renewable resources. Further implications incorporate undermining shared objectives and values of cultural diversity as well as the capacity to contribute to the community. The list further entails the dilution of norms, traditions, social compact, attachment to the land and national identity (Corroto, 2012). Indeed, attempts to alter norms and behaviours of the population's standard patterns of production and consumption, unless proven practically viable and affordable, shall face considerable risk of public resistance.

However, Australia's strategic structural advantages derive from its democratic orientation, rights and liberties, the rule of law, cultural diversity and economic development, including its world-class financial, touristic and educational industries (Commonwealth of Australia, 2014). The government's espoused goal is to promote inclusive economic growth by acting upon reducing the widening wealth gap as a means of improving social equity. It shall improve resource management and enforce environmental accountability through subsidies and mandatory periodical reporting, and shall set up regulations to control environmental degradation. The broad scope of the global mutual interdependence further dictates Australia to extend its hands to global efforts on sustainable development. Corporate social responsibility in Australia shall reduce ecological footprints and restore environmental damage. Australian enterprises operating overseas are to act as vehicles for sustainability, which requires these firms to redefine and integrate social and environmental goals to their profit targets and top executives' compensation packages. The government might introduce tax incentives for sustainable businesses operating within the Asia-Pacific region to promote its dedication towards sustainability. Acknowledged for its top ranking on civil liberties and freedom of association (The Economist Intelligence Unit, 2017), Australian community-driven NGOs retain the highest potential to lead the change towards sustainable development. This is through conducting a bottom-up approach that involves articulating a strategic vision, coordinating collective actions, informing stakeholders about the consequences of sustainability challenges, raising finance and planning emergency operations.

4.6 From GDP Growth-Based/Fossil Fuel-Based Developmental Model to Sustainable Development

In an attempt to dismantle the public misconception regarding the uses and limitations of GDP, this section conducts an initial overview of the governing wider context that nurtured and consolidated this misconception. This wider context comprises both the neoclassical approach to economics as well as the role of global neoliberalism in promoting GDP as an adequate measure of economic development.

The mainstream neoclassical approach advocates pursuing limitless and infinite economic growth. Additionally, it retains a tendency to disregard the political context and neglect social implications of poverty, inequality and prolonged unemployment accompanying its standardly advocated socioeconomic directions. The rationale behind its approach seems inherent in the endless wealth accumulation and exponential growth that defines capitalism; however, it is deemed impossible in a finite world (Aronson, Blignaut, Milton, & Clewell, 2006). This contradiction has brought both the neoclassical approach and global neoliberal capitalism under increasing scepticism and critique (Amin, 2012; Kothari et al., 2015). Indeed, the neoclassical approach and the global capitalist system have failed in pricing resources, environmental welfare, social ills and externalities associated with their economic activities. They failed in valuing social cohesion, as well as public goods and harms for people excluded from the market economy because of their incapacity to afford market rates (McIntosh, 2008; Speth, 2008). Scientific consensus already affirms that the modern fossil fuel-driven economic system involving deforestation and industrial production, agriculture and livestock farming, all linked to global neoliberal capitalism, comprise the leading causes of ecological damage. However, switching to clean energy alone shall not solve underlying structural deficiencies without readdressing the logic behind the demand for the ever-increasing extraction, production and consumption accompanying the pursuit of endless global economic GDP growth-based development (Anderson, 2000; Hickel, 2016; Raworth, 2017).

The line of argument rationally extends to question the GDP-based economic growth model as a proxy for a country's standard of living, progress, development, welfare or prosperity (Coyle, 2014; Dowrick & Quiggin, 1998; Fioramonti, 2017). GDP merely accounts for monetary transactions; it does not capture nonmonetary activities that enhance the quality of life or strengthen social cohesion (Davies, 2004). Moreover, GDP neglects variations in human and environmental capital as well as many productive activities such as household tasks, volunteer community services and informal economy dealings (Costanza, Hart, Talberth, & Posner, 2009; Dowrick & Quiggin, 1998; Mitchell, 2009). It disregards the quality and concentration of growth among citizens thus implicitly fosters waste and social fragmentation (Badreldin, 2015). Evaluating Australia's economic growth as the world's largest coal exporter, with its expansions in fossil fuel-extracting industries, particularly new coal mines and coal-fired power plants, should extend beyond their positive reflections on GDP indices. It is more economically appropriate to internalise the uncounted devastating environmental costs of monster cyclones and storms, changing rainfall patterns and crop failures into such evaluations (Aronson et al., 2006).

4.7 Conclusion

The primary objective of this chapter is to investigate structural impediments that hinder Australia and the Asia-Pacific region from achieving their Paris Agreement targets and consolidating sustainable development. While neoliberal globalisation has nurtured ecological damage and widespread poverty and wealth inequalities in a systematic manner, this chapter argues the accumulation and persistence of these structural deficiencies portend severe implications against the attainment of sustainability targets. The chapter concludes that only the equitable, more inclusive and democratic developmental regimes are capable of consolidating sustainable growth and development. It further anticipates an inevitable transformation towards comprehensive structural reformations of economic, political and social foundations so as to overcome structural impediments to sustainable development (Badreldin, 2015; Kotz, 2009). The articulation of a strategic sustainability design dictates progress towards a more inclusive economic model empowered by concepts and practices of a sustainable enterprise economy, corporate social responsibility and corporate citizenship (Anderson, 2000; McIntosh, 2008, 2009).

Primary policy direction involves limiting human–environment interventions, setting lower ecological footprint targets, preserving social equity, investing in energy eco-efficiency and introducing practical models for low-carbon economies, while inducing lifestyle and behavioural changes that encourage value creation, pro-environmental habits, thrift and social and environmental entrepreneurship (Siraj-Blatchford, 2009). Additionally, it involves supporting a social consciousness-driven private sector fashioned by social entrepreneurs and empowered by vibrant labour representation, community-driven NGOs and a progressive taxation system capable of correcting market failures (Anderson, 2000). In light of the current economic impediments associated with neoliberal globalisation, there is a need to create an alternative to the current capitalist orthodoxy by redefining economic progress to acknowledge quality of life, community solidarity and environmental well-being (Aronson et al., 2006; Dafermos et al., 2015; Kothari et al., 2015).

Unless incorporating parallel transformations towards a fully functioning (representative) democratic governance, prospects for sustainable development shall remain minimal (Badreldin, 2015). Where all political mechanisms conventionally entail some degree of socioeconomic prejudice, undemocratic forms of rule are deemed neither inclusive nor stabilising, hence they are unsustainable from the origin. While struggling to consolidate its ruling at the expense of the public, with the later acknowledged as the sole agent of importance for a democratic polity, undemocratic governance typically encloses an embedded tendency to immunise itself against political contestation, thereby hindering alternative more efficient authority options. Despite many macroeconomic indicators, which may misleadingly suggest undemocratic governance may perform well, commonly in the short or medium run, structural imbalances will definitely emerge, develop and accumulate over the longer terms. Nevertheless, these structural deficiencies do and will override any claims of either realised or promised development made by undemocratic regimes.

References

Acemoglu, D., & Robinson, J. A. (2001). A theory of political transitions. *American Economic Association, 91*(4), 938–963.

Alesina, A., & Perotti, R. (1996). Income distribution, political instability, and investment. *European Economic Review, 40*(6), 1203–1228.

Amin, G. (2015, 20 May). Kayfa Zaharet "Modet" Elkhaskhasa? Transl. (How has the «fashion» of privatization appeared?). *Al-Shorouk*. Retrieved from http://www.shorouknews.com/columns/view.aspx?cdate=19052015&id=085a539b-692d-4463-b6f3-60eca4cd87b2

Amin, S. (2012). Ending the crisis of capitalism or ending capitalism. In K. E. I. Smith (Ed.), *Sociology of globalization: Cultures, economies, and politics* (pp. 129–145). Boulder, CO: Westview Press.

Amos, S. K., Keiner, E., Proske, M., & Radtke, F.-O. (2002). Globalisation: Autonomy of education under siege? Shifting boundaries between politics, economy and education. *European Educational Research Journal, 1*(2), 193–213.

Anderson, R. (2000). Climbing mount sustainability. *Reflections: The SoL Journal, 1*(4), 6–12.

Anderson, K., & Bows, A. (2012). A new paradigm for climate change. *Nature Climate Change, 2*(9), 639–640.

Anderson, S., & Cavanagh, J. (2004). Design for corporate rule. In J. Cavanagh & J. Mander (Eds.), *Alternatives to economic globalization: A better world is possible* (pp. 32–54). San Francisco, CA: Berrett-Koehler.

Aronson, J., Blignaut, J. N., Milton, S. J., & Clewell, A. F. (2006). Natural capital: The limiting factor. *Ecological Engineering, 28*(1), 1–5.

Australian Treasury. (2015). *Intergenerational report: Australia in 2055.* Canberra, Australia: Commonwealth of Australia.

Ayeb, H., & Bush, R. (2014, Fall). *Small farmer uprisings and rural neglect in Egypt and Tunisia.* Middle East report. Retrieved July 12, 2017, from http://www.merip.org/mer/mer272/small-farmer-uprisings-rural-neglect-egypt-tunisia#.VC_dkQ5Z9sU.twitter

Badreldin, A. (2015). Energy crisis keeps Egypt on the wrong side of capitalism. *The Global Studies Journal, 8*(4), 1–18.

Bivens, J. (2011). *Failure by design: The story behind America's broken economy.* New York, NY: Cornell University Press.

Brysk, A. (2002). Introduction: Transnational threats and opportunities. In A. Brysk (Ed.), *Globalization and human rights* (pp. 1–16). Berkeley, CA: University of California Press.

Cadman, T., Eastwood, L., Michaelis, F. L.-C., Maraseni, T. N., Pittock, J., & Sarker, T. (2015). *The political economy of sustainable development: Policy instruments and market mechanisms.* Cheltenham, UK/Northampton, MA: Edward Elgar.

Castells, M. (2011). *The power of identity: The information age: Economy, society, and culture* (Vol. 2). Hoboken, NJ: John Wiley.

Christopher, L. (1995). *The revolt of the elites and the betrayal of democracy.* New York, NY: W.W. Norton.

Clifton, R., & Maughan, E. (Eds.). (2000). *The future of brands: Twenty-five visions.* New York, NY: New York University Press.

Commonwealth of Australia. (2014). *Australian citizenship: Our common bond.* Canberra: The National Communications Branch of the Department of Immigration and Border Protection.

Corroto, P. (2012, Jan 15). ENTREVISTA A JACQUES RANCIÈRE: "Hablar de crisis de la sociedad es culpar a sus víctimas". Transl. (Jacques Rancière Interview: "Democracy is not, to begin with, a form of State"). *Publico.* Retrieved July 12, 2017, from http://www.publico.es/culturas/416926/hablar-de-crisis-de-la-sociedad-es-culpar-a-sus-victimas

Costanza, R., Hart, M., Talberth, J., & Posner, S. (2009). *Beyond GDP: The need for new measures of progress* (N. 4, Trans.). Boston: Pardee Center for the Study of the Longer-Range Future. Retrieved from https://www.bu.edu/pardee/files/documents/PP-004-GDP.pdf

Coyle, D. (2014). *GDP: A brief but affectionate history.* Princeton, NJ: Princeton University Press.

Dafermos, Y., Galanis, G., & Nikolaidi, M. (2015). A new ecological macroeconomic model: Analysing the interactions between the ecosystem, the financial system and the macroeconomy. *New Economics Foundation.*

Davies, G. F. (2004). *Economia: New economic systems to empower people and support the living world.* Sydney: ABC Books.

Dean, M.M. (2009). *Governmentality: Power and Rule in Modern Society.* Second edn. Sage Publications Ltd, London, California, New Delhi and Singapore.

Dowrick, S., & Quiggin, J. (1998). Measures of economic activity and welfare: The uses and abuses of GDP. In CSIRO Wildlife and Ecology & R. Eckersley (Eds.), *Measuring progress: Is life getting better* (pp. 93–107). Melbourne: CSIRO.

Elkington, J., & Hartigan, P. (2013). *The power of unreasonable people: How social entrepreneurs create markets that change the world.* Boston, MA: Harvard Business Review Press.

Esposito, J. (2014). Capitalism and democracy: Oil and water. *The Global Studies Journal, 6*(3), 33–46.

Fioramonti, L. (2017). *The world after GDP: Politics, business and society in the post growth era.* Cambridge, UK: Polity Press.

Friedman, B. (2013). *Democracy ltd. how money and donations corrupted British politics.* London: Oneworld.

Gills, B., & Rocamora, J. (1992). Low intensity democracy. *Third World Quarterly, 13*(3), 501–523.

Harvey, D. (2007). Neoliberalism as creative destruction. *The Annals of the American Academy of Political and Social Science, 610*(1), 21–44.

Harvey, D. (2011). *The enigma of capital: And the crises of capitalism.* London: Profile Books.

Hedges, C. (2010). *Death of the liberal class.* New York, NY: Nation Books.

Herman, E. S., & Chomsky, N. (1988). *Manufacturing consent: The political economy of the mass media.* New York, NY: Pantheon Books.

Hickel, J. (2016, July 15). Clean energy won't save us – Only a new economic system can. *The Guardian.* Retrieved July 12, 2017, from https://www.theguardian.com/global-development-professionals-network/2016/jul/15/clean-energy-wont-save-us-economic-system-can

Kaboub, F. (2011). *Understanding and preventing financial instability: Post-Keynesian institutionalism and government as employer of last resort.* Financial instability and economic security after the Great Recession, pp. 77–92.

Kothari, A., Demaria, F., & Acosta, A. (2015, July 21). Sustainable development is failing but there are alternatives to capitalism. *The Guardian.* Retrieved July 12, 2017, from https://www.theguardian.com/sustainable-business/2015/jul/21/capitalism-alternatives-sustainable-development-failing

Kotz, D. M. (2009). The financial and economic crisis of 2008: A systemic crisis of neoliberal capitalism. *Review of Radical Political Economics, 41*(3), 305–317.

Kristof, N. (2014, July 23). An idiot's guide to inequality. *The New York Times.* Retrieved July 12, 2017, from http://www.nytimes.com/2014/07/24/opinion/nicholas-kristof-idiots-guide-to-inequality-piketty-capital.html?smid=fb-share&_r=0

Madeley, J. (2003). *A people's world: Alternatives to economic globalization.* London: Zed Books.

McGrew, A. (2013). Globalization and global politics. In J. Baylis, S. Smith, & P. Owens (Eds.), *The globalization of world politics: An introduction to international relations* (6th ed., pp. 15–34). Oxford: Oxford University Press.

McInnes, K. L., White, C. J., Haigh, I. D., Hemer, M. A., Hoeke, R. K., Holbrook, N. J., ... Walsh, K. J. (2016). Natural hazards in Australia: Sea level and coastal extremes. *Climatic Change, 139*, 1–15.

McIntosh, M. (2008). Editorial. *The Journal of Corporate Citizenship, 30*, 3–9.

McIntosh, M. (2009, May). What is a sustainable enterprise economy? *APCSE Thinking Aloud.*

Mészáros, I. (2012). Structural crisis needs structural change. *Monthly Review, 63*(10), 19–32.

Minio-Paluello, M. (2014 July 15). The violence of climate change in Egypt. *Jadaliyya.* Retrieved July 12, 2017, from http://www.jadaliyya.com/pages/index/18548/the-violence-of-climate-change-in-egypt?fb_action_ids=10152639998169750&fb_action_types=og.recommends&action_object_map=[659400330818714]&action_type_map=[%22og.recommends%22]&action_ref_map=[]#.U8ZbMAeCRaI.facebook

Mitchell, T. (2009). Carbon democracy. *Economy and Society, 38*(3), 399–432.

Piketty, T. (2014). *Capital in the twenty-first century.* Cambridge, MA: Harvard University Press.

Posner, R. A. (1997). Equality, wealth, and political stability. *Journal of Law, Economics, and Organization, 13*(2), 344–365.

Raworth, K. (2017). *Doughnut economics: Seven ways to think like a 21st-century economist.* White River Junction, VT: Chelsea Green.

Reich, R. B. (2016). *Saving capitalism: For the many, not the few.* New York, NY: Vintage.

Robinson, R. (2004). Neoliberalism and the future world: Markets and the end of politics. *Critical Asian Studies, 36*(3), 405–423.

Schäuble, W. (2017, 17 March). This is how we can make globalization work for everyone. The World Economic Forum. Retrieved July 12, 2017, from https://www.weforum.org/agenda/2017/03/this-is-how-we-can-make-globalization-work-for-everyone/?utm_content=buffer73c4f&utm_medium=social&utm_source=facebook.com&utm_campaign=buffer

Siraj-Blatchford, J. (2009). Editorial: Education for sustainable development in early childhood. *International Journal of Early Childhood, 41*(2), 9–22.

Smith, D. (2007, May 22). Australia's carbon dioxide emissions twice world rate. *The Sunday Morning Herald*. Retrieved July 12, 2017 from http://www.smh.com.au/news/environment/australias-greenhouse-emissions-twice-world-rate/2007/05/22/1179601374518.html

Speth, J. G. (2008). *The bridge at the edge of the world: Capitalism, the environment, and crossing from crisis to sustainability*. New Haven, CT: Yale University Press.

Stiglitz, J. (2013). *The price of inequality: How today's divided society endangers our future*. New York, NY: W. W. Norton.

Stiglitz, J. E. (2006). *Making globalization work*. New York, NY: WW Norton.

Swift, R. (2002). *The no-nonsense guide to democracy*. London: Verso.

The Economist Intelligence Unit. (2017). *Democracy index 2016: Revenge of the "deplorables"*. The Economist Intelligence Unit Limited.

Watts, M., & Mitchell, W. (2012). Full employment. In J. E. King (Ed.), *The Elgar companion to post Keynesian economics* (pp. 229–236). Cheltenham, UK/Northampton, MA: Edward Elgar.

Wayne Nafziger, E., & Auvinen, J. (2002). Economic development, inequality, war, and state violence. *World Development, 30*(2), 153–163.

Wolff, R. D. (2012). *Democracy at work: A cure for capitalism*. Chicago, IL: Haymarket books.

Part II
Strategies to Achieve Sustainable Economy and Zero Emissions Future

Chapter 5
The Role of Planning Laws and Development Control Systems in Reducing Greenhouse Gas Emissions: Analysis from New South Wales, Australia

Nari Sahukar

Abstract Integrating emissions reduction into planning and environmental laws is a crucial mechanism in helping sub-national states or provinces play a role in climate change mitigation measures. In states like New South Wales (NSW), the absence of an integrated approach to considering and reducing greenhouse gas emissions is a critical law and policy gap that needs to be addressed.

This chapter focuses on reducing greenhouse gas emissions (mitigation) and sets out how planning and development control systems can be part of the solution for achieving emissions reduction targets.

It highlights two major structural barriers to taking effective action to reduce emissions in NSW, Australia's most populous state. The first is the lack of legislated or binding greenhouse gas emissions reduction targets, supported by regulatory infrastructure or agency responsibility for reducing emissions. The second is the lack of integration between the need to reduce greenhouse gas emissions and the land-use planning system.

This chapter focuses on the *Environmental Planning and Assessment Act 1979* (NSW) because most greenhouse gas emissions from NSW are authorised by planning and development approvals (explicitly or otherwise). It considers six key stages of the NSW planning system that are relevant to greenhouse gas emissions reduction. Within each of these key stages, the chapter illustrates how greenhouse gas emissions are currently dealt with and how the law can be improved to help reduce emissions.

EDO NSW is an independent, non-government community legal centre established in 1985 to assist community members and groups to protect the environment through law. The centre is a member of EDOs of Australia, a network of independent environmental legal centres across Australia.

N. Sahukar (✉)
Environmental Defender's Office New South Wales (EDO NSW),
Sydney, NSW 2000, Australia
e-mail: nari.sahukar@edonsw.org.au

© Springer International Publishing AG 2018
M. Hossain et al. (eds.), *Pathways to a Sustainable Economy*,
DOI 10.1007/978-3-319-67702-6_5

Overall, the aim of this chapter is to make clear that planning and development agencies and decision-makers need stronger laws and guidance on achieving emissions control. It concludes with 14 recommendations to address this problem.

Keywords Development Controls • Emissions Reduction • Environmental Impact Assessment • Greenhouse Gas • Land-Use Planning • Mining • Mitigation • New South Wales, Australia (NSW) • Planning Laws • Strategic Planning

> *This chapter was first published as a report by EDO NSW (2016b), and revised for publication in this volume. The primary author is Nari Sahukar, Senior Solicitor, Policy & Law Reform team at EDO NSW, with editorial assistance gratefully received from Jeff Smith, Rachel Walmsley, Megan Kessler and Emma Carmody.*

5.1 Introduction

This chapter is about the need for sub-national leadership and climate change regulations, focusing on Australia's most populous state, New South Wales (NSW). It is argued that planning, environmental assessment and development controls (collectively, "the planning system") need to play a prominent role in reducing anthropogenic greenhouse gas emissions. This includes emissions from a wide range of sources that pass through planning and development approval, such as energy production, consumption and exports from NSW (NSW EPA, 2015).[1]

Australia and the wider Pacific region are already seeing the effects of 1 °C warming, and further climate change impacts are locked in by emissions already in the atmosphere (Australian Government BOM and CSIRO, 2014). Predicted impacts for NSW include:

- Up to 10 additional days above 40 °C each year in northern NSW by 2030, rising to 33 additional days by 2070
- Increased crop failure, human and animal deaths
- Longer and more intense bushfire seasons
- Accelerated biodiversity loss
- Increased irreversible soil erosion, affecting food security and water quality (NSW OEH, 2016; CSIRO, 2015).

The Intergovernmental Panel on Climate Change (IPCC, 2014) is highly confident that:

---

[1] NSW has one of the highest levels of greenhouse gas emissions per capita in the world, estimated at 19.6 tonnes per person per year (NSW EPA, 2015). This is well above Organisation for Economic Co-operation and Development (OECD) and global averages, but below the Australian average of 23.2 tonnes.

Without additional mitigation efforts beyond those in place today, and even with adaptation, warming by the end of the 21st century will lead to high to very high risk of severe, widespread, and irreversible impacts globally …

The Paris Agreement provides clear impetus for strong action and targets on climate change across government, business and community sectors (UNFCCC COP21, 2015).[2]

There are two broad ways to limit climate change impacts: mitigation and adaptation. Mitigation involves avoiding and reducing emissions while adaptation involves increasing resilience to unavoidable change. Mitigating emissions now reduces the damage and costs of adapting later.

This chapter focuses on mitigation and sets out how the planning system can be part of the solution to avoid the impacts of climate change. It highlights two major structural barriers to NSW taking effective action to reduce emissions.

The first is a lack of legislated or binding greenhouse gas emissions reduction targets, supported by regulatory infrastructure or agency responsibility for reducing greenhouse impacts. Notwithstanding a recent aspirational government objective to achieve "net-zero" emissions by 2050, it is clear there is no government agency responsible for reducing the state's greenhouse gas emissions, and no coordinated plans or laws to make it happen.[3]

NSW is one of the worst-performing Australian jurisdictions in this regard. South Australia, Tasmania, Victoria and the Australian Capital Territory have shown that specific emissions reduction targets are responsible and achievable (see Table 5.1). Other states and nations in Europe, the United Kingdom, North America and Asia are also legislating targets and actions to protect their climate.

The second barrier is the lack of integration between the need to reduce greenhouse gas emissions and the land-use planning system. State governments are mainly responsible for making laws for the environment, regional and urban planning, and natural resource management. However, no part of the NSW planning, development assessment, approval or licensing framework does any of the following:

- Performs strategic climate risk assessment
- Links to an emissions reduction target or a finite "carbon budget"[4]
- Has an overarching goal to avoid 2 °C warming[5]

---

[2] In December 2015, more than 190 nations affirmed a goal to reduce greenhouse gas emissions to limit average global warming to well below 2 °C above pre-industrial levels and to pursue efforts to limit warming to 1.5 °C. The Paris Agreement builds on past international commitments in Cancun, Lima and elsewhere under the 1992 UN Framework Convention on Climate Change.

[3] Based on analysis of legislation, case law and related discussions with regulatory agencies over several years.

[4] Carbon budgets, also known as "global emissions budgets", "have gained prominence as a way to analyse and communicate the scale of emissions reductions required to remain within a global temperature limit. Emissions budgets help to link emissions targets and trajectories to the underlying science of climate change" (Australian Government Climate Change Authority, 2014).

[5] While environmental impact statements are required to predict emissions from individual project proposals, we know of no policy stating how planning authorities take this into account, individually or cumulatively.

Table 5.1 Summary of State and Federal Climate Mitigation Laws and Targets

State	Climate mitigation law	Legislative targets
SA	*Climate Change and Greenhouse Emissions Reduction Act* 2007	• 60% reduction in emissions by 2050 (1990 baseline) • Net-zero emissions by 2050 (new target to be legislated)[a] • 50% of electricity generated from renewables by 2025 (policy target)
Tas	*Climate Change (State Action) Act* 2008	• 60% reduction in emissions by 2050 (1990 baseline)
ACT	*Climate Change and Greenhouse Gas Reduction Act* 2010	• Net-zero emissions by 2050 (principal target) • 40% reduction in emissions by 2020 (1990 baseline) • Peaking per capita emissions by 2013 • 100% of electricity generated from renewables by 2020
Vic	*Climate Change Act* 2017[b]	• Net-zero emissions by 2050 (principal target)[c] • Requires 5-yearly interim targets and climate strategies
Cth	(No legislated emissions reduction target) *Renewable Energy (Electricity) Act* 2010; *Carbon Credits (Carbon Farming Initiative) Act* 2011	• 5% reduction in emissions by 2020 (2000 baseline) (Kyoto Protocol, policy target) • 26–28% reduction in emissions by 2030 (2005 baseline) (Paris Agreement, policy target) • Revised Renewable Energy Target of 33,000 GWh by 2020 • *Note:* The Climate Change Authority (2015) recommended a 30% reduction in emissions by 2025 (2000 baseline), with a further target range of 40–60% reduction by 2030[d]
NSW	No climate mitigation law	• None • "Aspirational" objective of net-zero emissions by 2050[e]
NT	No climate mitigation law	• None
Qld	No climate mitigation law	• None
WA	No climate mitigation law	• None

[a]Source: Government of South Australia (2015)
[b]The *Climate Change Act* 2017 (Vic) received Royal Assent on 28 February 2017. It replaces the *Climate Change Act* 2010 (Vic)
[c]Source: Victorian Government Department of Environment, Land, Water and Planning (2017)
[d]Source: Fraser (2015)
[e]Source: NSW OEH (2016)

This chapter aims to make clear that decision-makers need stronger laws and guidance on greenhouse gas emissions reductions. The absence of an integrated approach to considering and reducing greenhouse gas emissions is a critical policy gap that needs to be filled.

The main planning law in NSW is the *Environmental Planning and Assessment Act* 1979. This chapter focuses on this Act because most of New South Wales' greenhouse gas emissions are authorised by planning and development approvals (explicitly or otherwise).

5 The Role of Planning Laws and Development Control Systems in Reducing…

In recent times, the NSW Government's position has been that the planning system is not the place to curb greenhouse gas emissions. This approach needs to change. Properly integrating climate change and emissions reduction into planning and environmental laws is a crucial mechanism in helping NSW play a role in mitigation measures.

5.2 Assessment and Critique of the NSW Planning Act

This chapter focuses on six key stages of the NSW planning system that are relevant to greenhouse gas emissions reduction:

1. Setting the framework
2. Strategic planning
3. Environmental impact assessment
4. Development decisions
5. Other approvals
6. Compliance and enforcement

Within each of these key stages, the chapter considers how climate change and greenhouse gas emissions are currently dealt with under the *Environmental Planning and Assessment Act* 1979 (NSW) (Planning Act), and then how the law can be improved to help reduce these emissions.

5.2.1 Setting the Framework

Tackling greenhouse gas emissions reduction in NSW requires consideration of two related issues: assigning responsibility within the government and designing effective planning laws, from high-level objects through to operations.

5.2.1.1 Assigning Responsibility

The Current Approach

There is no NSW Government agency responsible for reducing the state's greenhouse gas emissions, and no coordinated plans or laws to do so. Historically a leader in planning and environmental matters, NSW is currently lagging behind other jurisdictions in this regard. Table 5.1 demonstrates this point. Concerningly, in recent years the NSW Government has made the argument that the planning system is not the place to curb greenhouse gas emissions.[6]

In November 2016, the NSW Government announced an "aspirational" objective to achieve "net-zero" emissions by 2050. It also called for comment on a strategic plan to guide the existing NSW Climate Change Fund (a levy on power generators) from 2017 to 2022. The draft plan proposed actions to assist clean energy investment, energy efficiency and adaptation.

While the announcement of a statewide target of net-zero emissions by 2050 is welcome, any realistic chance of achieving it requires legal and institutional support. For example, the state of Victoria has also adopted a net-zero by 2050 target, but this will be implemented by replacing that state's *Climate Change Act* 2010 (Vic) with a new *Climate Change Act* 2017 and reinstating an emissions target.

5.2.1.2 The Way Forward

Robust climate change mitigation measures should require the NSW planning system to be informed by greenhouse gas emissions targets set under legislation, and to place duties on decision-makers to achieve those targets. This has been done in other countries and states.[7]

The NSW State Plan should include a vision for a rapid and responsible transition to a low-carbon economy in NSW. This is consistent with NSW membership of The Climate Group States & Regions Alliance. All government departments would need to direct and report on climate change action, including various agencies responsible for environmental management and land-use planning. Similar approaches are adopted in the *Environment (Wales) Act* 2015,[8] and recommended in the 2015 review of Victoria's *Climate Change Act* 2010 (Wilder, Skarbek and Lyster, 2015).

5.2.2 Planning Objectives

5.2.2.1 The Current Approach

The objects or purpose of an Act set out its aims, and they guide and interpret how the Act applies. In NSW, the Planning Act contains no reference to climate change or the need to reduce greenhouse gas emissions, either in its objects or in its operational provisions.

---

[6] As argued by the Department of Planning in the Ulan case, discussed below: see *Hunter Environment Lobby Inc v Minister for Planning* [2011] NSWLEC 221 (24 November 2011) at [59].

[7] For example, the *Climate Change (Scotland) Act* 2009 (Scot) places a legal duty on the Scottish Ministers to "ensure that the net Scottish emissions account for the year 2050 is at least 80% lower than the [1990] baseline". It also legislates an interim target for 2020, subject to expert advice.

[8] The *Environment (Wales) Act* 2015 (Wales) puts ecologically sustainable development at the heart of the Welsh natural resource management laws, including climate change (see Part 2 of the Act).

In the absence of an explicit object to this effect, climate change and greenhouse gas emissions have been dealt with in a more circuitous way.

At the development consent stage, decision-makers are required to consider the environmental, social and economic impacts of a development proposal and the public interest (Planning Act, s. 79C). The public interest, in turn, is informed by the objects of the Planning Act, which include "the protection of the environment" and "encouraging ecologically sustainable development". The Courts have interpreted that the public interest requires at least the high-level consideration of ecologically sustainable development[9] and its principles. This has included consideration of a project's impacts on climate change, and vice versa.[10]

While climate change considerations have therefore been read into the Planning Act, a clearer signal would be for the objects of the Planning Act to be explicit about reducing emissions and protecting NSW against climate change impacts.

5.2.2.2 The Way Forward

As Table 5.1 shows, climate mitigation laws in other states and territories set legislative targets for emissions reductions. Some also set renewable energy targets. This should be supported by whole-of-government policy that assigns specific responsibilities to achieve these goals.

An important next step is to embed emissions reduction in planning law objects.

Planning law in Queensland goes some way towards this. The overall purpose of the *Sustainable Planning Act* 2009 (Qld) is "to seek to achieve ecological sustainability" (s. 5). The Act provides guidance on how this is done, setting out an inclusive list of how to advance the Act's purpose. This includes:

- Ensuring decision-making processes "take account of short- and long-term environmental effects of development at local, regional, state and wider levels, including, for example, the effects of development on climate change"
- Prudent use of natural resources, including "considering alternatives to the use of non-renewable natural resources"

---

[9] In NSW planning, pollution and environmental laws, ecologically sustainable development calls for the integration of environmental, social and economic considerations in decisions, based on the principles of ecologically sustainable development. These derive from the *Protection of the Environment Administration Act* 1991 (NSW) section 6, and include:

- The precautionary principle (i.e. that scientific uncertainty should not delay action to avert serious harm)
- Conservation of biodiversity and ecological integrity as a fundamental consideration
- Intergenerational equity (and intra-generational equity)
- Full valuation of environmental costs and benefits (including the polluter pays principle).

[10] See for example *Walker v Minister for Planning* (2007) 157 LGERA 124; [2007] NSWLEC 741; *Minister for Planning v Walker* (2008) 161 LGERA 423; [2008] NSWCA 224; *Aldous v Greater Taree City Council* (2009) 167 LGERA 13; [2009] NSWLEC 17.

- "avoiding, if practicable, or otherwise lessening, adverse environmental effects of development, including, for example—climate change and urban congestion"

It is crucial that the objects of an Act are clearly operationalised in decision-making. Unlike the NSW Planning Act, the Queensland Act mandates that powers or functions are exercised in a way that advances the Act's purpose. However, this Act does not apply to all development in Queensland, and addressing climate change has not been systematically adopted in other relevant laws.[11]

The objects of the NSW Planning Act must refer explicitly to reducing greenhouse gas emissions as a key aim of the planning system. Importantly, legislation should also include specific duties for decision-makers, to ensure reduction targets are met (as in the UK).[12]

Greenhouse gas emissions reduction should then be explicitly referred to at significant decision-making points in the Planning Act. As discussed below, this would include strategic plan-making, environmental impact assessment, development approval, and compliance monitoring and reporting.

5.2.3 Strategic Planning

5.2.3.1 The Current Approach

Strategic planning is a high-level process for managing land use and natural resources. It should ensure that smaller development decisions accord with longer-term social, economic and environmental needs, and adjust to changing conditions.

The Planning Act deals with strategic planning.[13] Regional plans are now in development, but there is no clear legislative or policy framework to consider climate change and no systematic or practical measures to plan for a low-carbon future.[14] The planning system also remains disconnected from broader natural resource management planning.[15]

---

[11] This includes laws that regulate major projects with significant greenhouse gas emissions, such as the *State Development and Public Works Organisation Act* 1971 (Qld).

[12] For example, *Climate Change (Scotland) Act* 2009 (Scot) and *Environment (Wales) Act* 2015 (Wales).

[13] *Environmental Planning & Assessment Act* 1979 (NSW) Part 3.

[14] Regional plans are instead given effect by delegated ministerial directions under s 117 of the Act.

[15] For example, under the *Local Land Services Act* 2013 (NSW) or *Natural Resources Commission Act* 2003 (NSW).

Strategic Planning Principles

The government's Planning Review of 2011–2013 found widespread agreement that planning law needs much clearer requirements for strategic (or regional) planning. The Independent Planning Review Panel recommended that any new Act should set out clear objects for strategic planning (Moore & Dyer, 2012), including to:

> Consider the scientifically anticipated impact of climate change within the footprint of the strategic planning study area and the broad measures required to mitigate its impact.

While the government's Planning Bill 2013 proposed a new strategic planning framework, it did not accept the panel's recommendation to include climate change as a strategic planning factor.[16] To date, the NSW Government has not adopted this recommendation.

State Environmental Planning Policies

The Planning Act empowers the government to enact State Environmental Planning Policies (SEPPs) for a range of state-level matters. SEPPs are powerful instruments because they can override Local Environmental Plans (LEPs) and development controls, and have other pervasive effects.[17]

While SEPPs could be used to assess, limit or regulate greenhouse gas emissions through planning and development approvals, there is no currently comprehensive "climate change mitigation SEPP" that integrates these considerations into decision-making under the Planning Act.

There are many SEPPs dealing with development, including state significant development and infrastructure, smaller public infrastructure, mining and extractive industries (Mining SEPP), residential energy and water standards (BASIX) and environmental protection (such as for koala habitat, coastal rainforests and wetlands).[18] The Mining SEPP and BASIX are two SEPPs that do refer to greenhouse gas emissions, but only in limited ways.

---

[16] In fact, climate change, greenhouse gas emissions, urban sustainability and design were all surprisingly absent from the Government's Planning Green Paper (2012), White Paper (2013) and the draft Planning Bill 2013 (which ultimately stalled).

[17] SEPPs generally require public consultation before they are made, but they can have pervasive effects that override local development controls, and once a SEPP is made, these effects are difficult to challenge: *Huntlee Pty Ltd v Sweetwater Action Group Inc*; *Minister for Planning and Infrastructure v Sweetwater Action Group Inc* (2011) 185 LGERA 429; [2011] NSWCA 378.

[18] For example, SEPP 2011; SEPP (Infrastructure) 2007; SEPP (Mining, Petroleum Production and Extractive Industries) 2007; SEPP (Building and Sustainability Index) 2004; SEPP No 44—Koala Habitat Protection.

Resource and Infrastructure Projects

The lack of strategic planning and emissions reduction targets has meant that sectors with significant greenhouse gas impacts—such as mining, energy and transport—have expanded on an ad hoc, project-by-project basis. At present, over three-quarters of NSW emissions come from extracting, processing and burning fossil fuels, including transport emissions (now at 19%) (NSW OEH, 2016).[19] However, there remains no coordinated government plan to manage emissions-intensive sectors.

In recent years, the need to consider the extractive resources sector in a way that reduces land-use conflicts has led to various strategic planning frameworks.[20] Yet none of these frameworks deal with reducing greenhouse gas emissions. Neither the frameworks, nor the studies and reports that underpin them, properly acknowledge climate change and emissions reduction as key challenges that need to be dealt with by government, industry and the community.

For example, the Strategic Statement on NSW Coal (NSW Government, 2014a) includes a "sustainability" objective related to "triple bottom line considerations to promote comprehensive and balanced decision making". However, it omits any reference to considering climate change risks and impacts, or ecologically sustainable development principles such as the precautionary principle, intergenerational equity or full environmental costing.

Similarly, under the Strategic Release Framework (2015) for opening new areas to coal and gas exploration, greenhouse gas emissions and climate risks are excluded from "triple bottom line" considerations. The reason given for excluding these matters from assessments of new coal licence areas was that these are not local issues relevant to "preliminary regional impact assessment".[21] This approach fails to recognise the fact that the impacts of climate change will be felt both locally and globally in all sectors of society.

---

[19] See also State of the Environment 2015—Greenhouse Gas Emissions (NSW EPA, 2015, p. 37).

[20] These frameworks include:

- *Strategic Regional Land Use Policy* (2012): A government policy dealing with land and water use conflicts between agriculture and coal seam gas. These plans are only in place in the Upper Hunter and New England/North West.
- *Strategic Statement on NSW Coal* (2014): A government policy statement which "aims to realise… economic value while protecting our environment and the health of our communities".
- *NSW Gas Plan* (2014): A Government plan to "pause and reset" coal seam gas regulation (NSW Government 2014b).
- *Strategic Release Framework* (2015): A government framework for opening new areas for coal and coal seam gas exploration. The framework is part of the *NSW Gas Plan* and the government's response to various reports, including the NSW Independent Commission Against Corruption report into corruption in coal mining licensing (ICAC, 2013); the NSW Chief Scientist & Engineer's report into coal seam gas regulation (O'Kane, 2014); and the NSW Coal Exploration Steering Group (2015).
- *Minerals Industry Action Plan* (2015): An industry plan commissioned by government to support mining in NSW.

[21] The NSW Government Coal Exploration Steering Group (2014) stated that "the [strategic preliminary issues] assessment will not consider non-local issues such as the management of greenhouse gas emissions".

5.2.3.2 The Way Forward

Strategic planning in NSW needs to deal more effectively with greenhouse gas emissions reduction. This could be done through, among other things, implementing a reductions target through the Planning Act, which should in turn carry through to state-level planning policies (such as SEPPs), regional strategic plans and LEPs.

NSW strategic planning laws must require plan-makers to plan for the direct and indirect impacts of climate change—including the need to reduce greenhouse gas emissions—when developing and finalising strategic plans. In doing so, plan-makers must act in accordance with state and national greenhouse gas emissions reduction targets,[22] and the best available science.

Better planning is also needed for emissions-intensive sectors with long project lifecycles. Resource development and transport infrastructure must be ecologically sustainable, including consideration of climate change. This aim should be given upfront in the strategic planning phase. For example, having regard for international agreements to avoid 1.5–2 °C average global warming, the NSW Government should not release new areas for resource extraction that will significantly contribute to greenhouse gas emissions, and state and national emission reduction targets must be embedded in law and policy.

This would need new whole-of-government guidance and investment in resourcing, training and cultural change within the public and private sectors.

5.2.4 Environmental Impact Assessment

5.2.4.1 The Current Approach

Environmental impact assessment (EIA) relies on comprehensive and accurate information on the potential environmental, social and economic impacts of a development proposal. In NSW, this involves the proponent submitting an environmental study to the decision-maker. This is an important stage for considering greenhouse gas impacts, because most high-emitting projects require some form of EIA and approval in the planning system.

NSW planning laws set out two main EIA processes that broadly reflect the scale of development impacts.[23]

---

[22] This would include laws that give effect to international agreements to avoid dangerous climate change.

[23] *Environmental Planning and Assessment Act* 1979 (NSW) Parts 4 and 5.

Higher-impact proposals, such as large factories, mining production and transport projects, require a full environmental impact statement (EIS) prepared in accordance with Environmental Assessment Requirements issued by the Department of Planning.[24] The EIS is generally placed on public exhibition with the development proposal for at least 30 days' comment.

Lower-impact proposals, such as local development, infrastructure and mining and gas exploration, require a Statement of Environmental Effects or a Review of Environmental Factors (under Part 4 or 5 of the Planning Act). These processes are generally less likely to assess greenhouse gas emissions.

The EIA process does not adequately deal with emissions reductions for a number of reasons.

First, there is no standard legal provision, and no standard policy statement, on how the impacts of greenhouse gas emissions are to be assessed for particular sectors or project types.[25]

Recently there have been some limited steps to address this. For example, new requirements for major mining proposals will require the EIS to comprehensively forecast and assess their greenhouse gas emissions, including downstream or Scope 3 emissions, such as from burning exported coal (NSW Government, 2015). Also, recent economic assessment guidelines expect mining and coal seam gas companies to address the costs of greenhouse gas emissions, "including quantification where feasible" (NSW Government, 2015a). However, Scope 3 emissions were excluded from these guidelines, with detailed requirements deferred to future technical chapters. For other sectors, expectations for assessment remain unclear.

Second, there remains wide discretion and little guidance on what to do with emissions information once decision-makers have it (addressed in Sect. 5.2.5, below). Indeed, no project in NSW has been rejected on the basis of excessive greenhouse gas emissions or unacceptable risks to the climate, and there is no established framework for doing so.

Third, it is widely recognised that cumulative impact assessment is a key inadequacy of EIA in the NSW planning system.[26] Greenhouse gas emissions are the quintessential example of cumulative impacts because they incrementally add up to a shared and dangerous problem. As a result, local or state planning policies tend to dismiss greenhouse gas emissions as a "global issue", and individual assessments dismiss the significance of a project's emissions.

---

[24] These include "designated development", and most state significant development and infrastructure.

[25] The BASIX scheme for housing efficiency is the main exception, discussed further below.

[26] For example, the term *cumulative* is not used in the Planning Act and appears only three times in the *Environmental Planning and Assessment Regulation* 2000 (NSW) (cl 228 and Schedule 3—Designated development).

5.2.4.2 The Way Forward

Planning laws can greatly improve EIA processes by (1) standardising assessment requirements for high-emitting sectors and (2) requiring major project applications to include a climate impact statement to highlight greenhouse gas emissions and mitigation. In particular, they should:

- Clarify in law the scale of projects, impacts and sectors that are required to estimate greenhouse gas emissions in EIA documentation via a new climate impact statement
- Demonstrate how a project will avoid, minimise and offset emissions
- Require the use of standard methods to estimate direct Scope 1 emissions (such as fugitive methane from a coal mine), Scope 2 emissions (such as electricity use) and up-and-downstream Scope 3 emissions[27]
- Prescribe a method to calculate the full social costs of greenhouse gas emissions, including environmental and public health costs over time (see, for example, US EPA, 2015a)
- Require the EIA to estimate a range of emissions, the degree of any uncertainty and the reasons for such uncertainty; these assessments should be conducted by independent experts.

The NSW planning system also needs to consider a project's greenhouse gas emissions in the context of the contribution that NSW and Australia must make to keep global temperatures from rising more than 1.5 °C. Appropriate climate mitigation architecture, such as a carbon budget and emissions register, would substantially assist in this endeavour (see Recommendations 1 and 14).

5.2.5 Development Decisions

5.2.5.1 The Current Approach

This section considers the decision-making process and development consent conditions for projects involving significant greenhouse gas emissions.

---

[27] Standard methodologies have been developed for defining and accounting for Scope 1, 2 and 3 emissions. For example, the Indicative Standard Environment Assessment Requirements for mining projects (NSW Department of Planning and Environment, 2015b) adopt the Greenhouse Gas Protocol of the World Resource Institute (n.d.) and others.

Under the NSW Planning Act, most projects with significant greenhouse gas emissions are dealt with and approved under Part 4 and the matters for consideration outlined in section 79C(1).[28]

In deciding whether to approve or refuse a proposal, the decision-maker must evaluate the EIA information and associated reports, public submissions, relevant planning instruments (such as SEPPs and LEPs), likely impacts on the environment, suitability of the site and the public interest.[29] If approved, conditions may be imposed on the project to minimise adverse impacts, including on the environment.

NSW environmental and planning law has made some advances to deal with greenhouse gas emissions over the past decade, often through public interest litigation in the Courts. However, at various times, the NSW Government has resisted the rigorous consideration of greenhouse gas emissions that is increasingly needed and being adopted elsewhere. This resistance has been evident in landmark cases over the past decade, such as *Gray*, *Walker*, and *Ulan*,[30] and in recent departmental recommendations, major project approvals and conditions (EDO NSW, 2016a).

Approvals Under the Mining SEPP

For resource extraction proposals, the Mining SEPP further informs decision-making. This is a powerful planning instrument which in part determines where mining can take place in NSW.[31] The Mining SEPP does require decision-makers to consider greenhouse gas emissions from a project in two ways (while still being heavily geared to facilitate resource extraction).

The first is a duty to consider the need for conditions to minimise greenhouse gas emissions. Notably, it is not a duty to *impose* those conditions.[32] The second is a requirement to consider an assessment of the project's greenhouse gas emissions (including Scope 3 emissions). It is not clear *how* the decision-maker is to evaluate those emissions or weigh them against other factors. It is also unclear exactly what

---

[28] Part 4 of the Planning Act includes ordinary development and state significant development under Division 4.1. Exceptions outside Part 4 include former "Part 3A" major projects (most of which are already approved) and State Significant Infrastructure (approved with broad ministerial discretion under Part 5.1 of the Planning Act, similar to former Part 3A).

[29] Generally, the consent authority for significant projects is the Planning Minister or their delegate (such as the Planning Assessment Commission).

[30] *Gray v The Minister for Planning, Director-General of the Department of Planning and Centennial Hunter Pty Ltd* (2006) 152 LGERA 258; [2006] NSWLEC 720; *Walker v Minister for Planning* (2007) 157 LGERA 124; [2007] NSWLEC 741; *Minister for Planning v Walker* (2009) 161 LGERA 423; [2008] NSWCA 224; *Hunter Environment Lobby Inc v Minister for Planning* [2011] NSWLEC 221 (24 November 2011); *Hunter Environment Lobby Inc v Minister for Planning* [2012] NSWLEC 40 (13 March 2012).

[31] *State Environmental Planning Policy (Mining, Petroleum Production and Extractive Industries)* 2007. For example, the Mining SEPP permits exploration and mining in certain areas and explicitly overrides other local development controls (see e.g. clauses 5–8).

[32] For example, it is open to the decision-maker to decide that consent for the mining project need not include conditions that greenhouse gas emissions be "minimised to the greatest extent practicable": cl 14.

"state or national policies, programmes or guidelines" the decision-maker must have regard to (or how), given that a central problem identified in this chapter is the lack of such policies.[33]

In the absence of clearer duties, current terms used in the Mining SEPP and some project conditions, such as to minimise emissions "to the greatest extent practicable", remain ambiguous. This affects the quality of decisions, approval conditions and enforceability.

Similarly, there is no guidance on what would constitute an unacceptable greenhouse gas impact that should lead to the refusal of a proposal.

Inconsistent or Non-existent

Overall, current decision-making requirements to assess and limit greenhouse gas emissions are vague and inconsistent, if they exist at all. This is ultimately reflected in the high-level conditions placed on recent projects with significant greenhouse gas emissions. For example: "The Applicant shall (a) implement all reasonable and feasible measures to minimise the ...release of greenhouse gas emissions from the development ..." (NSW Government Department of Planning and Environment, 2015c).

The qualified nature of this requirement highlights the broad discretion left open to the proponent and the Department of Planning. This approach falls strikingly short of a requirement to avoid, minimise and offset all greenhouse gas emissions—a proposition partly advanced by the community before the Courts in the *Ulan* case of 2011–2012.[34]

> **Ulan Mine Expansion: Greenhouse Gas Emissions, Offsets and Approval Conditions**
> In *Ulan*, a local community group, the Hunter Environment Lobby, brought a merits review case in the NSW Land and Environment Court against a coal mine expansion. Among other things, the group sought conditions requiring the mining company to offset its Scope 1 and 2 greenhouse gas emissions.
>
> In its initial judgement, the Court noted its intention to impose the community group's suggested offset conditions, and sought the parties' comments on the implications of new federal clean energy legislation for the wording of those conditions (as carbon pricing was soon to commence).

---

[33] In future, the *Indicative SEARs* will guide the proponent's EIS preparation. However, we are not aware of any other formal guidelines that assist the decision-maker. By referring to "policies" and not laws, the SEPP highlights the absence of any specific State or federal law to reduce emissions.

[34] *Hunter Environment Lobby Inc v Minister for Planning* [2011] NSWLEC 221 (24 November 2011); *Hunter Environment Lobby Inc v Minister for Planning* [2012] NSWLEC 40 (13 March 2012).

> However, in its second judgement, the Court ultimately declined to impose the greenhouse gas conditions, on the basis that the federal carbon pricing regime would cover most of the mine's activities that result in Scope 1 emissions, and therefore the purpose of the condition would be otherwise met.
>
> Shortly after the second judgement, however, a change of Australian Government saw the repeal of the clean energy laws, which the Court had relied on to displace the need for greenhouse gas offset conditions. In effect, the coal mine was no longer required to pay for its greenhouse gas emissions, either under the federal scheme or the state development approval.
>
> Despite the federal repeal, the NSW Department of Planning (which opposed the greenhouse offset conditions) has not taken the Court's lead to impose greenhouse gas offset conditions in subsequent recommendations or approvals.
>
> *Ulan* is a landmark case in that the greenhouse gas conditions sought by the community group were the first of their kind to be considered by a Court in Australia. The first judgement—where the Court expressed an intention to impose greenhouse gas conditions, subject to consideration of the clean energy regime—sets an important precedent.

Building Sustainability Standards

BASIX is the primary mechanism for improving building sustainability standards in NSW law. The BASIX SEPP imposes energy and water efficiency standards and limits on new residential buildings and major alterations. In 2013, it was estimated that BASIX saved two million tons of greenhouse gas emissions (NSW Department of Planning and Infrastructure, 2013). However, the standards have not been updated between 2006 and 2016, and the majority of NSW energy-related emissions remain out of the scope of BASIX.

Significantly improving building sustainability standards is important because:

- Higher efficiency standards can create significant co-benefits, including consumer savings.
- Half of NSW greenhouse gas emissions are from stationary energy (e.g. electricity) (NSW OEH, 2016).
- A quarter of NSW emissions are from electricity use in housing (NSW EPA, 2015).
- More than half of Sydney's emissions come from the *non*-residential built environment.[35]

---

[35] Commercial and manufacturing/industrial sectors together account for 55% of Sydney's emissions (NSW Government Department of Planning and Infrastructure, 2012).

5.2.5.2 The Way forward

Quantifying total greenhouse gas emissions at the EIA stage and placing this information before the public and the decision-maker are important steps forward. However, decision-makers also need clear guidance on what to do with that information and how to weigh it up.

NSW planning laws must adopt a comprehensive assessment and decision-making framework for the climate change implications of development, particularly major projects. Key missing pieces include emission reduction targets linked to the planning system and guidelines that direct decision-makers to assess the significance of emissions, refuse unacceptable impacts and impose standardised conditions to avoid, minimise and offset greenhouse gas emissions.

Reforms are also needed to deal with high-emissions facilities that operate under long-outdated approvals and conditions. Planning and pollution laws, and the agencies that administer them, must impose updated standards. This could be done by updating development consent conditions and pollution licence conditions or through pollution reduction programmes overseen by the NSW Environment Protection Authority (EPA).[36]

For smaller-scale urban development, where the problem is cumulative impacts, there are a number of ways to improve upon the sound principles and good works of BASIX,[37] namely:

- Continually update building sustainability standards in light of technological developments, requiring regular reviews to stimulate continuous improvement and innovation
- Expand mandatory minimum efficiency standards to commercial and industrial buildings, including retrofitting, where significant gains could be made
- Allow planning authorities to set more stringent sustainability standards for precincts
- Use NSW expertise developed through BASIX to lead and develop national standards for other sustainability measures, such as lifecycle emissions and waste levels (EDO NSW, 2014)

---

[36] The *Protection of the Environment Operations Act* 1997 (NSW) requires the EPA to periodically review EPLs and empowers the EPA to set pollution reduction programs for licensed facilities: ss 68, 78. For development consents, the Planning Act, s 96, which enables modification at proponents' request, could be expanded.

[37] While the two million tons noted above is not insignificant, these savings remain a very small fraction of NSW emissions (less than half of 1% of annual emissions from stationary energy). EDO NSW (2014) estimated savings from BASIX at around 0.35% of the state's annual stationary energy emissions. Comparisons in 2009 estimated BASIX savings of 0.04% of annual NSW emissions (Thorpe & Graham 2009).

5.2.6 Other Approvals

5.2.6.1 The Current Approach

Some development proposals with significant greenhouse gas emissions and impacts require additional permits beyond development approval under the Planning Act. For example, this may include a pollution licence (an environment protection licence, EPL) or a mining title (exploration licence or production lease).

There is scope for these other authorisations to limit greenhouse gas emissions, set emissions standards,[38] extend existing "polluter pays" schemes[39] or require offsets.[40] However, in practice, they are generally not used in this way. As with planning laws, NSW mining and pollution laws fail to provide for adequate assessment, limitation or continuous improvement of the greenhouse gas emissions and climate change impacts of major projects.

5.2.6.2 The Way Forward

Pollution licences and mining titles are two of the most important environmental controls linked to the planning system (though they are administered under other laws and agencies).

In other jurisdictions, principles like "continuous improvement" and "best available technology" are used to keep environmental standards up to date (EDO NSW, 2012; EDOs of Australia, 2015). A similar approach could be adopted in NSW by imposing greenhouse gas emissions limits and load-based licensing fees on

---

[38] There are no mandatory greenhouse efficiency standards for power stations in NSW or Australia. Pollution laws are generally state-based, with limited (sometimes glacial) national coordination. By contrast, in 2015 the Obama Administration imposed national emissions standards via the United States Environmental Protection Agency's Clean Power Plan (US EPA 2015b) and the *Clean Air Act* (US). However, the Trump Administration is expected to dismantle this framework.

[39] Some pollutants in NSW are subject to load-based licensing, a "polluter pays" fee based on how much pollution is emitted by a licensed facility. However, the NSW pollution licensing system does not generally limit greenhouse gas emissions, or charge load-based licence fees on carbon dioxide or methane emissions. This is despite the fact that *fugitive emissions* from coal mining, oil and gas infrastructure account for approximately 11% of NSW greenhouse gas emissions (NSW EPA, 2015), on top of emissions from burning fossil fuels for electricity.

[40] EPLs already play an important role in regulating air pollution in NSW. These licences and other requirements regulate air, land and water pollution from major projects via the *Protection of the Environment Operations Act* 1997 (NSW) and related regulations.

pollution licences. Likewise, pollution licences or development consent conditions could be modified to impose requirements for emissions avoidance, minimisation or offsets, and for monitoring, auditing and reporting.

There are various ways to put this in practice:

- Via *pollution licence reviews*, which the EPA conducts at least every 5 years (or as a standard requirement when licensees seek to modify their licence)
- Via *development modifications*, as a standard requirement when existing operators apply to the Department of Planning to modify their development consents (as when mining companies seek to expand operations)
- Via *legislative amendments* to the Planning Act:
  – To *explicitly permit* the Department of Planning to *update consent conditions* and require "continuous improvement" of environmental standards and/or
  – To *explicitly require* decision-makers to *impose conditions* for greenhouse gas emissions standards, limits and offsets when granting or modifying consents

In addition, NSW resources law should include an object to develop resources only in a manner that is compatible with minimising greenhouse gas emissions (including fugitive and exported or Scope 3 emissions), and reducing NSW's cumulative contribution to climate change. For example, as noted, NSW should not release new areas for extraction that will significantly increase emissions, given international agreements to avoid 1.5–2 °C warming.

In concert with planning laws, applications for exploration and mining titles (and renewals) should also be required to address likely greenhouse gas emissions that a project will emit. Licensing considerations would be coordinated with EIA, climate impact statements and development approval considerations, to make decisions on the best and most up-to-date information, and to avoid unacceptable impacts early in the process.

5.2.7 Compliance and Enforcement

5.2.7.1 The Current Approach

There is insufficient focus on monitoring and enforcing compliance with development approvals and other licence conditions relating to greenhouse gas emissions. There has been a recent effort to improve the monitoring and enforcement capacity of the Department of Planning, but it is unclear if any additional resources are being directed to monitor and enforce conditions on greenhouse gas emissions. It is also unclear if there is sufficient data to do so effectively.

The lack of emission reduction targets, strategic planning or development standards and conditions in the Planning Act means there is no limit on greenhouse gas emissions approved under the planning system. There is also no public monitoring of total emissions approved.

5.2.7.2 The Way Forward

A best practice system for accountability and quality assurance would build in requirements to monitor, report on and improve emission levels from high-emitting development and quantify all cumulative emissions. This would include assembling and analysing information collected via consent conditions, regulators, independent auditors and public reporting.

A mandatory greenhouse gas monitoring, reporting and auditing register should also be established for individual facilities with significant carbon footprints in NSW (with costs recovered by an industry levy). This could monitor and publicly report on facility-level emissions, limits and compliance with development approvals for operations above a certain pollution threshold. It would draw on and supplement data published by the National Greenhouse and Energy Reporting Scheme.[41]

5.3 Conclusions and Recommendations for the NSW Planning System

5.3.1 Setting the Framework

Recommendation 1
Enact new climate change laws that include provisions that:

- Set a clear overarching objective to reduce greenhouse gas emissions
- Impose duties on government ministers to set periodic and long-term emissions reduction targets and carbon budgets, based on expert advice
- Where appropriate, set a legislative renewable energy target for electricity use
- Require the new Act's implementation and goal-setting to be consistent with internationally agreed climate goals, best available science and ecologically sustainable development principles.

Recommendation 2
Insert an objective in NSW planning law to reduce greenhouse gas emissions in accordance with those duties, targets, carbon budgets, global goals and best available science.

---

[41] NGERS publishes aggregate corporate data and limited facility-level reporting for Scope 1 and 2 emissions of certain power stations and high-emitting corporations (many of whom own multiple high-polluting facilities) (Australian Government Clean Energy Regulator 2016).

5.3.2 Strategic Planning

Recommendation 3
Amend NSW planning law to require that strategic plans contribute to reducing, monitoring and improving greenhouse gas emissions across sectors, in accordance with relevant targets and best available science. Update all state environmental planning policies (SEPPs) accordingly.

Recommendation 4
Before releasing a new resource exploration or extraction area, require the relevant minister to consider:

- Likely emissions from resultant projects in the context of drawing down a state or national carbon budget
- The scale, cost and timing of lifecycle greenhouse gas emissions of a project
- Potential cumulative impacts with other past, present and future (approved or proposed) projects

5.3.3 Environmental Impact Assessment

Recommendation 5
Require consistent and independent assessment of the likely greenhouse gas emissions of all major projects. This must include a climate impact statement that states:

- How the project proposal contributes to relevant goals and targets to reduce greenhouse gas emissions
- Specific measures to avoid, minimise and offset emissions from the project
- The measures in place to ensure downstream emissions are avoided, minimised and offset
- The full cost of the project's emissions
- Full and proper consideration of alternative options

Recommendation 6
Publish greenhouse gas assessment and decision-making guidelines to ensure consistent, robust assessment and decisions based on best available science. Guidelines should apply an "avoid, mitigate and offset" hierarchy for reducing emissions.

Recommendation 7
Require mandatory accreditation of environmental consultants who prepare environmental impact assessment reports and independent appointment of those consultants to projects.

5.3.4 Development Decisions

Recommendation 8
Strengthen decision-making requirements for development approvals and conditions in the Planning Act, with the aim of achieving emissions reduction targets. In particular, establish new duties to:

- Have regard to state and national emissions trajectories and act in accordance with short- and long-term reduction targets
- Consider the level of greenhouse gas emissions as grounds for refusal of (or a duty to refuse) unacceptable impacts
- Impose specific conditions on development consents and mining titles to minimise emissions, meet certain standards if the project is approved, and offset emissions that cannot be minimised or avoided
- Apply clear guidelines, rules and standards to minimise and offset emissions.

Recommendation 9
Amend NSW planning laws to clarify that development consent conditions can be updated to require continuously improved standards, whether or not a modification has been requested.

Recommendation 10
Expand the Building Sustainability Index for energy and water efficiency standards (or an equivalent system) by:

- Including significantly higher residential efficiency standards
- Expanding mandatory efficiency standards to commercial and industrial buildings
- Building in review periods that require standards to continuously improve
- Leading and developing national standards for other sustainability measures such as lifecycle emissions and waste levels.

5.3.5 Other Approvals and Licences

Recommendation 11
Mandate climate change and emissions as a consideration for assessing exploration or production title applications under mining laws. Before issuing a mining title, the relevant minister should be required to consider:

- Likely emissions in the context of drawing down a state or national carbon budget
- The scale, cost and timing of lifecycle greenhouse gas emissions of a project
- Cumulative impacts with other past, present and approved projects.

Recommendation 12
Add greenhouse gases as pollutants in NSW pollution control laws, to recognise their contribution to environmental degradation and encourage behavioural change. In the absence of a carbon price, this should include load-based licensing fees for greenhouse gas emissions, consistent with the polluter pays principle.

Recommendation 13
Establish emissions standards and continuous improvement requirements for NSW power stations, where appropriate, based on nationally consistent standards. Standards and requirements would be enforceable conditions on pollution licences.

5.3.6 Compliance and Enforcement

Recommendation 14
Establish a comprehensive greenhouse gas monitoring and auditing register to report on individual facilities with significant carbon footprints in NSW. This would draw on existing and new data, to track and report on approved and actual emissions.

References

Australian Government Bureau of Meteorology (BOM) and Commonwealth Scientific and Industrial Research Organisation (CSIRO). (2014). *State of the climate 2014*. Retrieved from http://www.bom.gov.au/state-of-the-climate

Australian Government Clean Energy Regulator. (2016). *Greenhouse and energy information 2014–15*. Retrieved from http://www.cleanenergyregulator.gov.au/NGER/Published-information/Reported-greenhouse-and-energy-information-by-year/greenhouse-and-energy-information-2014-15

Australian Government Climate Change Authority. (2014). *Reducing Australia's greenhouse gas emissions: Targets and progress review—final report*. Retrieved from http://www.climatechangeauthority.gov.au/reviews/targets-and-progress-review-3

CSIRO. (2015). *New climate change projections for Australia*. Retrieved from http://www.csiro.au/en/News/News-releases/2015/New-climate-change-projections-for-Australia

EDO NSW. (2012). *Clearing the air: Opportunities for improved regulation of pollution in NSW*. Retrieved from http://www.edonsw.org.au/clearing_the_air_opportunities_for_improved_regulation_of_pollution_in_new_south_wales

EDO NSW. (2014). *Submission on the Building Sustainability Index (BASIX) target review*. Retrieved from http://www.edonsw.org.au/submission_on_the_building_sustainability_index_basix_target_review

EDO NSW. (2016a). *Past cases—Climate change and energy*. Retrieved from http://www.edonsw.org.au/climate_change_energy_cases

EDO NSW. (2016b). *Planning for climate change: How the New South Wales planning system can better tackle greenhouse gas emissions*. Retrieved from http://www.edonsw.org.au/planning_for_climate_change

EDOs of Australia. (2015). *Submission to the National Clean Air Agreement*. Retrieved from http://www.edonsw.org.au/anedo_submission_to_the_national_clean_air_agreement

Fraser, B. (2015). *Some observations on Australia's post-2020 emissions reduction target*. Retrieved from http://www.climatechangeauthority.gov.au/sites/prod.climatechangeauthority.gov.au/files/files/CFI/CCA-statement-on-Australias-2030-target.pdf

Government of South Australia. (2015). *South Australia's climate change strategy 2015–2050—Towards a low carbon economy*. Retrieved from https://www.environment.sa.gov.au/Science/Science_research/climate-change/climate-change-initiatives-in-south-australia/sa-climate-change-strategy

ICAC (Independent Commission Against Corruption). (2013). *Investigation into the conduct of Ian Macdonald, Edward Obeid Senior, Moses Obeid and others*. Retrieved from http://www.icac.nsw.gov.au

IPCC (Intergovernmental Panel on Climate Change). (2014). *Climate change 2014: Synthesis report. Contribution of Working Groups I, II and III to the Fifth Assessment Report of the Intergovernmental Panel on Climate Change*. Retrieved from http://www.ipcc.ch/

Moore, T., & Dyer, R. (2012). *The way ahead for planning in NSW* (Independent Review Panel Report. Vol. 1).

NSW EPA (Environment Protection Authority). (2015). *State of the environment 2015*.

NSW Government. (2012). *Strategic regional land use policy*. Retrieved from http://www.resourcesandenergy.nsw.gov.au/landholders-and-community/coal-seam-gas/codes-and-policies/strategic-regional-land-use-policy

NSW Government. (2014a). *Strategic statement on NSW coal*. Retrieved from http://www.resourcesandenergy.nsw.gov.au/__data/assets/pdf_file/0006/521637/Strategic-statement-on-NSW-coal.pdf

NSW Government. (2014b). *NSW gas plan*. Retrieved from http://www.resourcesandenergy.nsw.gov.au/energy-supply-industry/legislation-and-policy/nsw-gas-plan

NSW Government. (2015). *Strategic release framework*. Retrieved from http://www.resourcesandenergy.nsw.gov.au/__data/assets/pdf_file/0006/521637/Strategic-statement-on-NSW-coal.pdf

NSW Government Coal Exploration Steering Group. (2014). *Improving NSW's process to allocate coal exploration licences*. Retrieved from http://www.resourcesandenergy.nsw.gov.au/__data/assets/pdf_file/0008/534158/CESG-Stakeholder-Forum-Issues-Brief-Consultation-Document.pdf

NSW Government Department of Planning and Environment. (2015a). *Guidelines for the economic assessment of mining and coal seam gas proposals*. Retrieved from http://www.planning.nsw.gov.au/Policy-and-Legislation/Mining-and-Resources/~/media/C34250AF72674275836541CD48CBEC49.ashx

NSW Government Department of Planning and Environment. (2015b). *Indicative secretary's environmental assessment requirements for state significant mining proposals*. Retrieved from http://www.planning.nsw.gov.au/Policy-and-Legislation/Mining-and-Resources/~/media/6A2B386AFC324ECA9B4FFD0BC5D3AF20.ashx

NSW Government Department of Planning and Environment. (2015c). *Warkworth Continuation Project (SSD-6464)—Development consent under Section 89E of the Environmental Planning and Assessment Act 1979*.

NSW Government Department of Planning and Infrastructure. (2012). *Sydney over the next 20 years—A discussion paper*. Retrieved from http://catalogue.nla.gov.au/Record/5980840

NSW Government Department of Planning and Infrastructure. (2013). *BASIX target review—FAQs*. Retrieved from https://www.basix.nsw.gov.au/iframe/images/4050pdfs/BASIX-Target-Review-GeneralQA_extension.pdf

NSW Government Minerals Industry Taskforce. (2015). *Minerals industry action plan*. Retrieved from http://www.resourcesandenergy.nsw.gov.au/miners-and-explorers/programs-and-initiatives/miap

NSW OEH (Government Office of Environment and Heritage). (2016). *AdaptNSW*. Retrieved from http://climatechange.environment.nsw.gov.au

O'Kane, M. (2014). *Final report of the independent review into coal seam gas activities in NSW*. Retrieved from http://www.chiefscientist.nsw.gov.au/reports/coal-seam-gas-review

Thorpe, A., & Graham, K. (2009). Green buildings—Are codes, standards and targets sufficient drivers of sustainability in NSW? *Environment and Planning Law Journal, 26*, 486–497.

United Nations Framework Convention on Climate ChangeConference of the Parties 21 (UNFCCC COP 21). (2015). *Adoption of the Paris Agreement. Annex—Paris Agreement, Article 2* (FCCC/CP/2015/L.9/Rev.1). Retrieved from https://unfccc.int/resource/docs/2015/cop21/eng/l09r01.pdf

US EPA (United States Environmental Protection Agency). (2015a). *Social cost of carbon*. Retrieved from https://www.epa.gov/climatechange/social-cost-carbon

US EPA (United States Environmental Protection Agency). (2015b). *Overview of the Clean Air Act and Air Pollution Share*. Retrieved from https://www.epa.gov/clean-air-act-overview

Victorian Government Department of Environment, Land, Water and Planning. (2017). Retrieved from http://www.delwp.vic.gov.au/environment-and-wildlife/climate-change

Wilder, M., Skarbek, A. and Lyster, R. (2015). *Independent Review of the Climate Change Act 2010*. Retrieved from http://www.delwp.vic.gov.au/environment-and-wildlife/climate-change/2015-review-of-climate-change-act

World Resource Institute. (n.d.). *Greenhouse gas protocol*. Retrieved from http://www.ghgprotocol.org/

Case Law and Legislation Cited

Aldous v Greater Taree City Council. (2009) 167 LGERA 13.
Carbon Credits (Carbon Farming Initiative) Act 2011 (Cth).
Clean Air Act (US).
Climate Change (Scotland) Act 2009 (Scot).
Climate Change (State Action) Act 2008 (Tas).
Climate Change Act 2010 (Vic).
Climate Change Act 2017 (Vic).
Climate Change and Greenhouse Emissions Reduction Act 2007 (SA).
Climate Change and Greenhouse Gas Reduction Act 2010 (ACT).
Environment (Wales) Act 2015 (Wales).
Environmental Planning and Assessment Act 1979 (NSW).
Environmental Planning and Assessment Regulation 2000 (NSW).
Gray v The Minister for Planning, Director-General of the Department of Planning and Centennial Hunter Pty Ltd. (2006). 152 LGERA 258.
Hunter Environment Lobby Inc v Minister for Planning . (2011). NSWLEC 221 (24 November 2011).
Hunter Environment Lobby Inc v Minister for Planning. (2012) NSWLEC 40 (13 March 2012).
Huntlee Pty Ltd v Sweetwater Action Group Inc; Minister for Planning and Infrastructure v Sweetwater Action Group Inc. (2011). 185 LGERA 429.
Local Land Services Act 2013 (NSW).
Minister for Planning v Walker. (2008). 161 LGERA 423.
Planning Bill 2013 (NSW).
Protection of the Environment Administration Act 1991 (NSW).
Protection of the Environment Operations Act 1997 (NSW).
Renewable Energy (Electricity) Act 2010 (Cth).
State Development and Public Works Organisation Act 1971 (Qld).

State Environmental Planning Policy (Building and Sustainability Index). (2004).
State Environmental Planning Policy (Infrastructure) 2007.
State Environmental Planning Policy (Mining, Petroleum Production and Extractive Industries) 2007.
State Environmental Planning Policy (State and Regional Development) 2011.
State Environmental Planning Policy No 44—Koala Habitat Protection.
Sustainable Planning Act 2009 (Qld).
Walker v Minister for Planning. (2007). 157 LGERA 124.

Chapter 6
Carbon Disclosure Strategies in the Global Logistics Industry: Similarities and Differences in Carbon Measurement and Reporting

David M. Herold and Ki-Hoon Lee

Abstract The transportation industry can be regarded as a significant contributor to greenhouse gases, which puts pressure on global logistics companies to disclose their carbon emissions in the form of carbon reports. However, the majority of these reports show differences in the measurement and reporting of carbon information. Through an institutional theory lens, this chapter discusses the emergence and institutionalisation of carbon disclosure and their influence on carbon reporting, using the cases of FedEx and UPS. In particular, the chapter examines whether the carbon reports follow a symbolic or substantial approach and which institutional logic dominates the rationale behind each company's carbon disclosure. We uncover significant variances in the measurement and reporting of carbon performance information between those companies that provide insights into the different dominant logics and their influence on carbon reporting. Our results show that both companies adopt different dominant logics due to heterogeneous carbon reporting practices. In a quest for market-controlled sustainability initiatives, this research is important (a) for understanding the ways in which major carbon contributors operate between market and sustainability logics, and (b) for policymakers to provide an understanding of how to encourage investments in carbon performance without reducing the ability to compete in the marketplace, thus initiating a shift to more environmentally friendly activities.

Keywords Carbon disclosure • Sustainability reporting • Environmental sustainability • CO_2 • Institutional logics • Sustainability logic • Logistics • FedEx • UPS

D.M. Herold (✉) • K.-H. Lee
Griffith Business School, Griffith University, Queensland, Australia
e-mail: david.herold@griffithuni.edu.au

6.1 Introduction

Recent years are characterised by an increasing societal and scientific debate about climate change that often centres on corporate and industrial activities (Abbasi & Nilsson, 2016; Herold & Lee, 2017; Lee & Vachon, 2016). As a consequence, carbon disclosure initiatives have emerged at the international level and put pressure on companies to report their efforts and performance regarding greenhouse gas emissions (Kolk, Levy, & Pinkse, 2008). As the global logistics industry accounts for around 5.5% of global carbon emissions, carbon measurement and reporting has been increasingly adopted within the global logistics industry (KPMG, 2014; World Economic Forum, 2012). However, carbon measurement and reporting still represents a mainly voluntary organisational practice, and companies can choose which tools or guidelines to apply in order to measure sustainability and environmental performance (Hahn, Reimsbach, & Schiemann, 2015). As such, the carbon disclosure of global logistics companies shows a certain level of similarities and differences in the measurement and reporting of carbon performance information. Against this background, this chapter aims to examine the similarities and differences of the carbon reports, to understand the rationale behind each company's carbon disclosure behaviour.

In particular, we argue that the differences are attributed to different legitimisation approaches, which lead to either symbolic or substantial carbon reporting behaviour. In addition, the specific nature of the disclosure response may vary depending on emergence of what we call "sustainability logic". While companies are driven by the need for legitimation through the logic of sustainability or carbon reporting, they are also responsible for the well-being of their organisation at the same time, which is characterised by a market logic. Thus, we further argue that the reporting behaviour indicates the rationale or the dominant institutional logic behind the emergence and institutionalisation of carbon reports.

In this chapter, we therefore seek to find some answers about how the emergence of a sustainability logic influences the disclosure behaviour of carbon reporting. Using a qualitative research approach embedded within a comparative case study, the carbon information provided by FedEx and UPS through sustainability and Carbon Disclosure Project (CDP) reports will be analysed and discussed. After an introduction to carbon reporting in the global logistics industry with a focus on FedEx and UPS, the chapter discusses the emergence of carbon reporting from an institutional theory perspective. This is followed by an overview of different organisational approaches and the process of institutionalisation of carbon disclosure. The results of the analysis are then discussed and we conclude with recommendations for further research.

6.2 Carbon Disclosure in the Global Logistics Industry

Carbon disclosure represents a mainly voluntary organisational practice within companies (Hahn et al., 2015). By bypassing formal regulatory mechanisms, carbon reporting can be regarded as a "non-state market driven governance system" (Cashore, Auld, & Newsom, 2004, p. 4) or as a form of "civil regulation" (Murphy & Bendell, 1999). Its voluntary nature allows companies and management to choose between different measurement and reporting approaches to measure carbon emissions.

Both FedEx and UPS follow the Greenhouse Gas Protocol: A Corporate Accounting and Reporting Standard provided by the World Business Council for Sustainable Development and the World Resources Institute in 2004 (WRI/WBCSD, 2011). Within that guideline, differences in measuring and reporting between FedEx and UPS can be observed, although their operations can be considered to be very similar. Both companies are fully integrated across the four main transport modes: air, rail, road and ocean (Onghena, Meersman, & Van de Voorde, 2014). In addition, both companies are dominant players in the express business service, being able to service the majority of the world within 48 h (FedEx, 2014; UPS, 2014). The resources and systems of both exemplars are extensive in terminals, means of transportation, handling equipment, etc. To a large extent, the global network is based on air transport, mainly using their own aircraft, which ensures high quality and speed, i.e. time is an extremely important factor (Hertz & Alfredsson, 2003). The extensive use of the express network and its related carbon output has led to an integration of climate change into the business strategy of both companies.

The "Greenhouse Gas Protocol" allows companies to measure and report carbon emissions differently. A closer look at the carbon reports of FedEx and UPS reveals similarities, but also differences in the measurement and reporting of carbon emissions. To measure the targets, FedEx and UPS have implemented "intensity" targets rather than "absolute" targets. An "absolute" target would reduce the total amount of carbon emissions, and management fears that this would constrain the companies' growth. Therefore, an "intensity" target is preferred, which measures the target is a decline in carbon emissions relative to the level of logistical activity (McKinnon & Piecyk, 2012). However, companies need to decide which variables or indicators will be measured. Although FedEx and UPS have adopted the "intensity" approach, their variables/indicators differ. FedEx addresses only Scope 1 emissions and defines its carbon reduction target in "ton-miles" for their aircraft emissions and "miles per gallon" for their vehicle fleet (CDP, 2015b). While UPS addresses Scope 1 aircraft emissions also in "ton-miles", it has developed more detailed ratios for Scope 1 and 2 emissions of its divisions, with each assigned a different weighting factor (CDP, 2015c; McKinnon & Piecyk, 2012).

Differences can also be observed in the organisational structure of both companies and in how carbon disclosure in the broader context of sustainability is integrated into each company's strategy. UPS has appointed a Chief Sustainability Officer who has direct reporting responsibility to the UPS Management Committee

(CDP, 2015c). The Chief Sustainability Officer is responsible for the leadership of sustainable business practices as well as for meeting the emissions reduction goals. In contrast, FedEx delegates the management of environmental performance to the operating companies in line with business needs and has implemented Sustainability Impact Teams. These teams report to the FedEx Enterprise Sustainability Council, which is chaired by the Vice President for Environmental Affairs and Sustainability.

More importantly, FedEx and UPS use different measurement approaches. Within the Greenhouse Gas Protocol, companies can choose different carbon measurement and reporting schemes for their carbon emissions (WRI/WBCSD, 2011). These different schemes provide guidelines to set boundaries for carbon emissions reporting. Companies can choose between two different control approaches: "financial" or "operational" control. In both wholly owned and joint operations, the choice of approach changes how carbon emissions are categorised when operational boundaries are set, and therefore influences the amount of carbon emissions to be reported (WRI/WBCSD, 2011). While FedEx adopts the financial approach, UPS is adopting the operational approach (CDP 2015b, 2015c).

Furthermore, the extent to which Scope 3 emissions are disclosed is another point of difference within the global logistics industry. For global logistics companies, for instance, purchased transportation by air, rail, road and ocean accounts for the largest source of Scope 3 emissions (FedEx, 2014; UPS, 2014). This leads to uncertainty: as subcontractors usually do not disclose information on fuel burn, Scope 3 emissions are largely based on complex calculation models and scenarios taking into account data from operational systems such as origins, destinations and routing. Moreover, based on the data from CDP (2015a), only UPS seems to report their full Scope 3 emissions from purchased transportation. FedEx acknowledges the relevance of this data, but is not yet able to calculate it fully, as it contains data from the feeder aircraft contractors and the freight subdivision in the USA and Canada, but not from pick-up and delivery subcontractors outside the USA (CDP, 2015b).

Furthermore, when carbon reporting is examined in a logistics context, not only CO_2 emissions have to be taken into account, but all relevant greenhouse gas emissions (European Commission, 2007; Schmidt, 2009). Apart from CO_2, others include methane, nitrous oxide, sulphur hexafluoride, perfluorinated compounds and hydrofluorocarbons. The relevant indicator to measure the impact of emissions on climate change is CO_2 equivalents (Schaltegger & Csutora, 2012). However, variances in emission type reporting can be found within the cases of FedEx and UPS. While UPS reports four emissions types, FedEx reports only three.

It can be summarised that carbon reporting between FedEx and UPS is characterised by similarities and differences. While both companies seem to take carbon reporting seriously, significant differences in measurement and reporting can be observed. In the next section, we argue that these differences can be attributed to the influence of competing logics which leads, in turn, to different reporting approaches.

6.2.1 The Implications of Carbon Disclosure in the Global Logistics Industry

Climate change can be regarded as one of the biggest challenges faced by our planet and contemporary society. Environmental challenges are perceived as a threat to existing business models and companies are under scrutiny from various stakeholders (Kolk et al., 2008). In response to these stakeholder pressures, the disclosure of sustainability information about a company's activities in the form of sustainability reports has increased markedly over the last several years (Hahn et al., 2015; KMPG, 2014).

The emergence of carbon reporting can be attributed to the role of "institutional entrepreneurs" (DiMaggio, 1988). Institutional entrepreneurs attempt to transform the company's structures and norms by a "political process that reflects the power and interests of organized actors" (Maguire, Hardy, & Lawrence, 2004, p. 658). But even powerful institutional entrepreneurs cannot simply introduce new organisational practices within companies. Institutional entrepreneurs have to lobby for a change and create and build legitimacy in the organisational field to get approval from other actors for organisational action (Beckert, 1999). However, institutional theory is based on the view that action is not a choice among unlimited possibilities but, rather, presents a choice between a specific defined set of legitimate options (Wooten & Hoffman, 2008). As and when the organisational practice has been legitimised within a field, the organisation itself changes and institutionalisation has set in (Selznick, 2011).

Once an organisational practice such as carbon disclosure has been institutionalised in one company, other companies facing similar institutional pressures will eventually adopt similar strategies to gain legitimacy (DiMaggio & Powell, 1983; Scott, 1991; Suchman, 1995; Thornton, Ocasio, & Lounsbury, 2012). In other words, the institutionalisation of organisational practices provides the organising principles for a field and reflects the "assumptions and values, usually implicit, about how to interpret organisational reality, what constitutes appropriate behaviour and how to succeed" (Thornton, 2004, p. 70). These discursive practices represent what is called an institutional logic in the literature. Institutional logics underpin the appropriateness of organisational practices in given settings and at particular historical moments, which is influenced by multilevel political, cultural and social aspects of organisational behaviour and phenomena (Lounsbury & Ventresca, 2003). The emergence of sustainability in the global logistics industry can be regarded as the emergence of sustainability logic, as heightened concerns about corporate carbon emissions have created a potential legitimacy gap. To close this gap, powerful political contestants in the organisational field fight for the institutionalisation of the sustainability logic in the form of carbon reports (Suddaby & Greenwood, 2005).

While it can be acknowledged that corporate sustainability (or carbon disclosure) is the logic behind the search for legitimacy, companies are also driven by the logic of the market (Greenwood, Díaz, Li, & Lorente, 2010). Market logic assumes that companies address sustainability issues (only) if this positively affects their financial performance, such as profits or shareholder value (Schaltegger & Hörisch, 2015). Managers are constantly challenged to deal with sustainability while at the same time being responsible for the well-being of their organisation. Thus, the market logic and the sustainability logic can be regarded as coexisting, but competing, logics (e.g. D'Aunno, Sutton, & Price, 1991; Hoffman, 2004; Reay & Hinings, 2005). Consequently, organisational responses to their contexts are unlikely to be uniform and depend on the relative dominance of the respective logic.

These conflicting views and logics have a direct influence on the carbon reporting approach. Hrasky (2011) and Kim, Bach, and Clelland (2007) distinguish between a symbolic management approach and a substantial management approach. Symbolic behaviour in carbon reports may be rhetorical statements designed to create an impression of sustainable or environmental responsibility, which is not necessarily accompanied by corporate action (Hrasky, 2011; Kim et al., 2007). Symbolic behaviour can also be related to reputation management, which Schaltegger and Burritt (2015) describe as a company's focus on societal, political and media attention. In a symbolic or reputational approach, carbon-related activities and their reporting are closely linked to the public relations department to gain the support of the company's most immediate audiences (Hrasky, 2011). Moreover, symbolic management can be regarded as self-interested or narcissist behaviour in carbon reporting, with more or less substantiated claims of carbon-related achievements (Schaltegger & Burritt, 2015). In contrast, carbon reporting may reflect the substantial corporate action taken by a company to achieve carbon-related accomplishments, such as reducing its carbon footprint (Hrasky, 2011).

As such, the conflicting logics have an impact on the concrete design and reporting approach of the carbon report, particularly, whether a market or sustainability logic is dominant or whether a mix of both rationales can be found, which is mainly characterised by interactions between the two rationales. If carbon reporting is influenced by a market logic, a company will search for sustainability activities that increase revenue or reduce carbon emissions to save costs and link them to the finance and accounting department (Figge, 2005). Schaltegger and Burritt (2015) call this behaviour as the "business case" for sustainability, i.e. the identification and realisation of economic potential of voluntary environmental activities. The action-oriented or substantial approach links the information provided about carbon emission to a rather substantial reporting approach. If, however, a sustainability logic best describes the underlying rationale of carbon disclosure, a company will rather put the public relations and communications department in charge and consider different management activities as relevant, such as external communication of environmental and social activities, and apply approaches such as stakeholder dialogue or social and cultural sponsoring (Castelo Branco, Eugénio, & Ribeiro, 2008; Hogan & Lodhia, 2011; Michelon, 2011; Sridhar, 2012). This behaviour can be linked to a more symbolic reporting approach.

Based on the previous discussion, we argue that the conflict between the market and the sustainability logic may lead to opposing carbon reporting approaches between FedEx and UPS. Thus, the research aim of this chapter is to examine how the emergence of sustainability influences the disclosure behaviour of carbon reporting at FedEx and UPS. An examination of carbon reports of both companies may provide insights into the respective carbon disclosure behaviour, which leads to the following two research questions:

1. Which disclosure behaviour—symbolic or substantial—dominates the carbon reporting in the global logistics industry?
2. How does the emergence of sustainability influence the extent of carbon reporting at FedEx and UPS?

6.3 Methods

The research questions are examined through analysis of carbon reporting information provided by FedEx and UPS. This information will be drawn from two types of reports. First, carbon-related information is provided in the sustainability reports of both companies. These sustainability reports, the Sustainability Report at UPS and the Global Citizenship Report at FedEx, are publicly available and have been downloaded and printed, where available, for the time frame of 2005–2014. Second, the carbon information in the sustainability reports will be complemented by the latest information provided by FedEx and UPS in the reports of the CDP in 2015.

CDP is a prominent international collaboration that represents a voluntary effort to provide information relevant to investors on a wide range of climate-related activities, including measurement of emissions, organisational preparations, technological investments, and trading and offsets. Membership of the CDP was obtained to gain access to and obtain the carbon information that both companies provided to investors and specific stakeholders.

To assess the research questions, measures of specific disclosures related to issues associated with carbon-related information are needed. Although the logic or rationale behind carbon reporting is essential, this chapter focuses on the identification of specific carbon-related disclosures and not environmental or sustainability disclosures more generally. This is because if the purpose of disclosure is a legitimation approach in response to these specific carbon-related concerns, the response should be clearly identifiable with this area of concern. However, carbon-related information is linked to broader concerns related to greenhouse gas emissions and climate change. Thus, disclosures that contain, or are linked with, information about carbon-related issues are included in the analysis.

To answer the first research question, we focus on the latest sustainability and CDP reports and use four categories to capture disclosures that are reflective of a symbolic or substantial reporting approach. The first category deals with the company's carbon or carbon-related mission statement and how it is integrated and reflected in the business strategy. Both CDP and sustainability reports cover that information. The second category relates to the internal corporate initiatives to lighten the carbon footprint, while the third category relates to involvement in external initiatives to achieve a similar end. The fourth category deals with the degree of transparency and sheds lights on how clearly and understandably the carbon information is presented and displayed.

To compare and analyse the reports, we follow the approach of Styhre (2011) and adopt discourse analysis for the interpretation of these four categories. According to the definition of Du Gay (2000, p. 67), "a discourse … is a group of statements that provide a way of talking about and acting upon a particular object. When statements about an object or topic are made from within a certain discourse, that discourse makes it possible to construct that object in a particular way". In other words, discourse analysis is the contextualisation of communication and seeks to reveal the characteristics and motivations behind a text or the choice of words. In an organisational context, Hardy, Palmer, and Phillips (2000, p. 1235) argue that discourse theory does not suggest that "the 'realities' of the social world reside inside people's minds" but conceive of social relations as being embedded in social relations and identities that are previously "constituted in discourse and reified into institutions and practices". Thus, discourse analysis provides a foundation to reveal the meaning behind the statements in corporate reporting and whether FedEx or UPS follows a symbolic or substantial approach, or a mix of both.

To assess the second research question, a comparative analysis of disclosure behaviour over time is required. The data were collected manually from sustainability reports, where available, between 2005 and 2014 and three additional categories, five, six and seven, were used to analyse the influence of the sustainability logic in each company and in comparison. The fifth category consists of data that show how frequently and regularly sustainability reports and carbon information are provided to the stakeholders. In the sixth category, data are reported as an absolute and relative number of pages of carbon reporting in the sustainability report. This information shows the development of the extent of carbon reporting and indicates the importance of carbon-related information within the industry. The seventh and final category is a comparison of the extent of the reporting of Scope 3 emissions. This is important, as purchased transportation accounts for the largest source of Scope 3 emissions within FedEx and UPS. An analysis of these three categories provides insight into the influence of the emerging sustainability logic and how it affects the carbon report of each company.

6.4 Results and Discussion

Overall, the results show significant variances between the carbon reports of FedEx and UPS. It needs to be emphasised that the results provide insight about reporting behaviour and the emergence of sustainability and carbon reporting. It is not a quantitative comparison of carbon emissions, goals and achievements. We analyse the content of the sustainability and CDP reports and try to reveal the meaning behind the information and corporate statements of FedEx and UPS with regard to their carbon reporting and engagement.

A first indicator to analyse the commitment and the climate change strategy are corporate sustainability statements. Sustainability or climate change statements within a corporate report can be regarded as a reflection of the corporate strategy and a call for action through the introduction or implementation of an organisational practice. From an organisational perspective, corporate statements can be related to the concept of the "institutional statement", which Crawford and Ostrom (1995, p. 583) describe as "a shared linguistic constraint or opportunity that prescribes, permits, or advises actions or outcomes for actors (both individual and corporate)". In particular, a statement not only specifies the purpose, the conditions or sanctions, but also whether the statement must, must not or may be followed. If followed, an exclusively "must" set of institutional rules would maximise accurate reproduction, while a set of "must not" rules would predict only the boundaries of what is doable. However, a "may" set of institutional rules would neither demand nor prohibit a particular behaviour, which consequently leads to heterogeneity of action.

The heterogeneity is reflected in the different statements about sustainability and carbon emissions in the respective carbon reports of FedEx and UPS. The statement of UPS includes an overall goal ("to improve energy efficiency and to reduce greenhouse gas emissions") as well as a commitment to a proactive role "beyond complying with applicable laws" of corporate action to reduce carbon emissions (UPS, 2015). FedEx's approach to environmental sustainability is different. While FedEx also shows a commitment to reduce carbon emissions, the company delegates corporate action to all operating companies to "manage their environmental performance in line with business needs" (CDP, 2015b). This statement indicates a focus on the economic benefits of sustainability activities and is referred to by Schaltegger and Burritt (2015) as the "business case" of and for sustainability.

Both statements provide an indication of the rationale behind the carbon reporting. If sustainability activities are designed in a way that they will, or are likely to, increase profits, a company is likely to follow a profit-seeking rationale (Schaltegger & Hörisch, 2015). In contrast, an engagement "beyond complying with applicable laws" (UPS, 2015) indicates that sustainability activities are designed to secure legitimacy. Moreover, unlike FedEx, the UPS sustainability report includes a specific Climate Change Statement, which can be regarded as a rhetorical strategy to build legitimacy. According to Suddaby and Greenwood (2005), the development of a statement can build legitimacy if it includes institutional vocabularies that articulate the rationale behind an organisational practice and a language that reflects the pace and the necessity of change within the field.

The analysis indicates that both companies are committed to reduce carbon reporting, but follow different strategies. While FedEx emphasises a more profit-seeking behaviour of sustainability in their statement, UPS emphasises the legitimacy perspective to gain a competitive advantage over FedEx and other competitors.

The commitment to reduce the carbon footprint within the operational network is also reflected in the internal activities of FedEx and UPS. It needs to be emphasised that the internal activities in both companies are mainly related to operational excellence. In other words, carbon footprint reductions are directly linked to improving operational efficiency and fuel savings. Both companies divide their initiatives into an aircraft and a vehicle segment. To reduce aircraft emissions, which represent almost 80% of all transportation fleet emissions, FedEx established a Fuel Sense programme, which includes algorithms to better predict fuel consumption for departure and arrival of planes as well as to reduce weight during the flight (CDP, 2015b). UPS internal activities include the instalment of "winglets" to reduce fuel consumption. For the transportation fleet, FedEx developed a programme called Reduce, Replace and Revolutionize to improve vehicle efficiency, which mainly focuses on optimising routes and the use of electric vehicles. In comparison, UPS focuses on the implementation of an alternative fuel fleet that includes liquefied natural gas tractors and building-related infrastructure in the form of liquefied natural gas fuelling stations (CDP, 2015c). All these internal activities from both companies are voluntary initiatives and can be regarded as substantial activities to reduce the carbon footprint.

The external activities of both companies are very similar. FedEx and UPS both believe in a long-term shift to alternative or biofuels and support organisations that work on sustainable jet fuels. Moreover, both companies work together with universities and engage directly with policymakers and government agencies to lobby for standardised carbon-related measurement systems. Moreover, FedEx and UPS have representatives on the various boards of trade association to address climate change legislation. All these initiatives can also be regarded as substantial behaviour.

From a transparency point of view, the analysis shows differences between FedEx and UPS. While both companies provided extensive information about the Scope emissions and goals as well as about their internal and external activities, a second look reveals differences in both the sustainability and CDP reports. To measure carbon performance, companies can set different organisational boundaries. That is, they can follow different control approaches to measure carbon emissions, namely, the financial or the operational approach. While FedEx follows the financial approach, UPS applies the operational approach. In the financial approach, companies need only to report emissions from ventures in which they hold more than a 50% interest (WRI/WBCSD, 2011). In other words, FedEx does not need to report carbon from partnerships and cooperation for pick-up and delivery services if they do not own more than 50% of the partner's company. Compared to UPS, "this approach may lead to less complete reporting", because when an operational control is applied, the carbon measurement and reporting "is not limited to majority-held ventures, it also applies to minority ventures" (IPIECA, 2011, pp. 3–5). Thus, an operational approach can be regarded as a more complete and transparent approach, as it tries to capture emissions from the entire operational network.

Other indicators also tend to support more transparent behaviour of UPS over FedEx. FedEx refuses to provide details of carbon measurements and evaluation models for "competitive reasons" (CDP, 2015b), while UPS explains the background and the measurement factors and details. Moreover, UPS reports four of the seven greenhouse gases covered by the Kyoto Protocol, including a table with a conversion rate of "global warming potential", while FedEx reports only three greenhouse gases and does not provide any specific information (FedEx, 2014;

UPS, 2015). UPS also provides a detailed overview of the operational boundaries and a breakdown of carbon emissions per source and country areas, while FedEx does not seem to be able or willing to report their carbon output in detail (CDP, 2015b, 2015c).

Overall, FedEx and UPS have implemented substantial activities to reduce carbon emissions, which indicate the opposite of symbolic reporting behaviour. However, it seems that UPS started earlier to integrate carbon measurements into their strategy, which could explain the more detailed and comprehensive availability of carbon data at UPS as well as the more transparent information provided to stakeholders.

The earlier measurement and reporting of carbon emissions is also reflected in the frequency of carbon reporting at UPS; they consistently produced a full sustainability report annually from 2005 to 2014. In contrast, FedEx issued its first sustainability report in 2008 and the second one in 2010. In 2011 and 2012, FedEx provided updates to the sustainability report and returned only in 2013 and 2014 with a full sustainability report about the company's sustainability and carbon-related activities. UPS is also ahead when it comes to an emphasis on carbon reporting in the sustainability reports. Both companies have increased their information gradually between 2005 and 2014, but while FedEx provides carbon information on approximately four pages on average in their sustainability reports, UPS covers approximately nine pages on average (FedEx, 2014; UPS, 2015).

The analysis of Scope 3 emissions reporting shows a similar picture. As both companies use partners worldwide for pick-up and delivery services, purchased transportation by air, rail, road and ocean accounts for the largest source of Scope 3 emissions at FedEx and UPS. While UPS is able to calculate all relevant Scope 3 emissions and provide details of scope and boundaries, FedEx seems to be unable to calculate the relevant data. Their reporting of carbon emission related to pick-up and delivery services is limited to feeders from the Express Division as well as the fuel used by FedEx Ground and FedEx Freight in Canada (CDP, 2015b, 2015c). Moreover, UPS has verified Scope 1, 2 and 3 emissions by an independent auditor, while FedEx has verification for Scope 1 emissions only (CDP, 2014).

Several key findings can be drawn from the analysis of the categories and the comparison between FedEx and UPS. The analysis indicates that UPS was first to recognise the emergence of a legitimacy gap resulting from heightened concerns about carbon emissions and transform it into an organisational practice in the form of carbon reporting. From an institutional theory perspective, the emergence of carbon reporting at UPS can be linked to institutional entrepreneurs within UPS. In other words, the emergence of carbon reporting at UPS is a "process that reflects the power and interests of organized actors" (Maguire et al., 2004, p. 658) in attempting to institutionalise new organisational practices. These efforts to change or transform a field can stimulate social actions, by which "entrenched, field-wide authority is collectively challenged and restructured" (Rao et al., 2000, p. 276) and the organisational field itself becomes contested. Thus, the institutionalisation of climate change-related actions and carbon reporting can be regarded as a disruptive event, where UPS developed a new dynamic in the organisational field. In contrast, FedEx showed isomorphic behaviour, which can be attributed to either mimetic process or normative pressure. In particular, the institutionalisation of carbon reporting at UPS led to change in corporate practices at FedEx, to incorporate a stronger focus on sustainability and environmental concerns and, consequently, to the first publication of a sustainability report ("Global Citizenship Report") in 2008.

6.5 Conclusion

It is evident that both companies pursue different strategies with regard to sustainability and carbon reporting. UPS sees sustainability and carbon reporting as core activities and as an opportunity to gain a strategic advantage over competitors by exemplary behaviour. In contrast, while FedEx has also acknowledged the importance of sustainability and carbon reporting, sustainability activities to reduce the carbon footprint are driven by a "business-case" approach. In other words, action will only be taken if it is in line with business needs and an economic benefit can be generated. Although both companies are far from demonstrating symbolic behaviour, the analysis indicates that UPS shows more transparent and exemplary behaviour.

From an institutional theory perspective, we analysed the emergence of sustainability and carbon reporting and their influence on the organisational field. The analysis clearly shows that UPS was first to acknowledge the emergence of carbon emissions concerns and the struggle of institutional entrepreneurs within UPS to institutionalise carbon reporting. The institutionalisation of carbon reporting led to a change in the organisational field and to the emergence of a sustainability logic. As a result of the institutionalisation of the new logic, and in line with the traditional view of institutional theory, FedEx showed isomorphic behaviour and issued its own sustainability report at a later stage. We also found that the emergence of a sustainability logic has led to a gradual increase in carbon information in both companies. However, analysis of the carbon information leads to the conclusion that the carbon reporting of FedEx is rather dominated by a market logic that emphasises the economic benefits of sustainability, while UPS has prioritised the sustainability logic to gain a competitive advantage.

Besides these implications, this study faces some limitations. The case study comprises only two cases and is of qualitative nature, so further research in this area could find ways to quantify the carbon information in carbon reports and use a bigger sample to compare companies' carbon activities. Moreover, the cases are limited to the USA, so expanding the geographical scope might provide insights into cultural differences. For policymakers, this study provides a first step to understand why and how companies engage in carbon disclosure. However, it is still not clear if, for instance, specific carbon disclosure practices have an effect on carbon emissions, e.g. what kind of carbon disclosure practices lead to an actual reduction in carbon emissions. Future research may find ways to classify carbon disclosure behaviour and link it to actual carbon emissions reductions to identify the carbon disclosure drivers behind environmentally friendly activities.

References

Abbasi, M., & Nilsson, F. (2016). Developing environmentally sustainable logistics: Exploring themes and challenges from a logistics service providers' perspective. *Transportation Research Part D: Transport and Environment, 46*, 273–283.

Beckert, J. (1999). Agency, entrepreneurs, and institutional change. The role of strategic choice and institutionalized practices in organizations. *Organization Studies, 20*(5), 777–799.

Cashore, B., Auld, G., & Newsom, D. (2004). *Governing through markets: Regulating forestry through non-state environmental governance*. New Haven, CT: Yale University Press.

Castelo Branco, M., Eugénio, T., & Ribeiro, J. (2008). Environmental disclosure in response to public perception of environmental threats: The case of co-incineration in Portugal. *Journal of Communication Management, 12*(2), 136–151.

CDP. (2014). *Climate action and profitability: CDP S&P 500 climate change report 2014*. New York, NY: CDP North America.

CDP. (2015a). *Carbon disclosure project 2015*. Retrieved from http://www.cdproject.net/index.asp

CDP. (2015b). Investor CDP 2015 information request. In *Carbon disclosure project*. London: FedEx Corporation. Retrieved from https://www.cdp.net/en.

CDP. (2015c). Investor CDP 2015 information request. In *Carbon disclosure project*. London: UPS. Retrieved from https://www.cdp.net/en.

Crawford, S. E., & Ostrom, E. (1995). A grammar of institutions. *American Political Science Review, 89*(03), 582–600.

D'Aunno, T., Sutton, R. I., & Price, R. H. (1991). Isomorphism and external support in conflicting institutional environments: A study of drug abuse treatment units. *Academy of Management Journal, 34*(3), 636–661.

DiMaggio, P. J. (1988). Interest and agency in institutional theory. In L. G. Zucker (Ed.), *Institutional patterns and organizations: Culture and environment* (Vol. 1, pp. 3–22). Cambridge, MA: Ballinger.

DiMaggio, P. J., & Powell, W. W. (1983). The iron cage revisited: Institutional isomorphism and collective rationality in organizational fields. *American Sociological Review, 48*(2), 147–160.

Du Gay, P. (2000). Markets and meanings: Re-imagining organizational life. In M. Schultz, M. Hatch, & M. Larsen (Eds.), *The expressive organization: Linking identity, reputation, and the corporate brand* (pp. 66–76). Oxford: Oxford University Press.

European Commission. (2007). *Carbon footprint: What it is and how to measure it*. Geneva: European Commission.

FedEx. (2014). *FedEx annual report*. Memphis, TN: FedEx Corporation.

Figge, F. (2005). Value-based environmental management. From environmental shareholder value to environmental option value. *Corporate Social Responsibility and Environmental Management, 12*(1), 19–30.

Greenwood, R., Díaz, A. M., Li, S. X., & Lorente, J. C. (2010). The multiplicity of institutional logics and the heterogeneity of organizational responses. *Organization Science, 21*(2), 521–539.

Hahn, R., Reimsbach, D., & Schiemann, F. (2015). Organizations, climate change, and transparency: Reviewing the literature on carbon disclosure. *Organization & Environment, 28*(1), 80–102.

Hardy, C., Palmer, I., & Phillips, N. (2000). Discourse as a strategic resource. *Human Relations, 53*(9), 1227–1248.

Herold, D. M., & Lee, K.-H. (2017). Carbon management in the logistics and transportation sector: An overview and new research directions. *Carbon Management, 8*(1), 79–97.

Hertz, S., & Alfredsson, M. (2003). Strategic development of third party logistics providers. *Industrial Marketing Management, 32*(2), 139–149.

Hoffman, A. J. (2004). *Climate change strategy: The business logic behind voluntary greenhouse gas reductions*. Ross School of Business paper.

Hogan, J., & Lodhia, S. (2011). Sustainability reporting and reputation risk management: An Australian case study. *International Journal of Accounting & Information Management, 19*(3), 267–287.

Hrasky, S. (2011). Carbon footprints and legitimation strategies: Symbolism or action? *Accounting, Auditing & Accountability Journal, 25*(1), 174–198.

IPIECA. (2011). *Petroleum industry guidelines for reporting greenhouse gas emissions*. London: International Petroleum Industry Environmental Conservation Association.

Kim, J.-N., Bach, S. B., & Clelland, I. J. (2007). Symbolic or behavioral management? Corporate reputation in high-emission industries. *Corporate Reputation Review, 10*(2), 77–98.

KMPG. (2014). *Corporate sustainability*. A progress report. Amsterdam: KMPG. Retrieved from http://www.sustainabilityexchange.ac.uk/files/corporate-sustainability-a-progress-report_1.pdf

Kolk, A., Levy, D., & Pinkse, J. (2008). Corporate responses in an emerging climate regime: The institutionalization and commensuration of carbon disclosure. *European Accounting Review, 17*(4), 719–745.

Lee, K. H., & Vachon, S. (2016). *Business value and sustainability: An integrated supply network perspective*. London: Palgrave Macmillan.

Lounsbury, M., & Ventresca, M. (2003). The new structuralism in organizational theory. *Organization, 10*(3), 457–480.

Maguire, S., Hardy, C., & Lawrence, T. B. (2004). Institutional entrepreneurship in emerging fields: HIV/AIDS treatment advocacy in Canada. *Academy of Management Journal, 47*(5), 657–679.

McKinnon, A., & Piecyk, M. (2012). Setting targets for reducing carbon emissions from logistics: Current practice and guiding principles. *Carbon Management, 3*(6), 629–639.

Michelon, G. (2011). Sustainability disclosure and reputation: A comparative study. *Corporate Reputation Review, 14*(2), 79–96.

Murphy, D. F., & Bendell, J. (1999). *Partners in time? Business, NGOs and sustainable development*. Discussion paper, United Nations Research Institute for Social Development (Vol. 109, pp. 1–71). New York: United Nations Research Institute for Social Development.

Onghena, E., Meersman, H., & Van de Voorde, E. (2014). A translog cost function of the integrated air freight business: The case of FedEx and UPS. *Transportation Research Part A: Policy and Practice, 62*, 81–97.

Rao, H., Morrill, C., & Zald, M.N. (2000). Power plays: How social movements and collective action create new organizational forms. *Research in Organizational Behavior, 22*, 237–281.

Reay, T., & Hinings, C. B. (2005). The recomposition of an organizational field: Health care in Alberta. *Organization Studies, 26*(3), 351–384.

Schaltegger, S., & Burritt, R. (2015). Business cases and corporate engagement with sustainability: Differentiating ethical motivations. *Journal of Business Ethics*, 1–19. doi:https://doi.org/10.1007/s10551-015-2938-0.

Schaltegger, S., & Csutora, M. (2012). Carbon accounting for sustainability and management. Status quo and challenges. *Journal of Cleaner Production, 36*, 1–16.

Schaltegger, S., & Hörisch, J. (2015). In search of the dominant rationale in sustainability management: Legitimacy-or profit-seeking? *Journal of Business Ethics*, 1–18. doi:https://doi.org/10.1007/s10551-015-2854-3.

Schmidt, M. (2009). Carbon accounting and carbon footprint—More than just diced results? *International Journal of Climate Change Strategies and Management, 1*(1), 19–30.

Scott, W. R. (1991). Unpacking institutional arguments. In W. W. Powell & P. J. DiMaggio (Eds.), *The new institutionalism in organizational analysis* (pp. 164–182). Chicago, IL: University of Chicago Press.

Selznick, P. (2011). *Leadership in administration: A sociological interpretation*. New Orleans, LA: Quid Pro Books.

Sridhar, K. (2012). Corporate conceptions of triple bottom line reporting: An empirical analysis into the signs and symbols driving this fashionable framework. *Social Responsibility Journal, 8*(3), 312–326.

Styhre, A. (2011). Competing institutional logics in the biopharmaceutical industry: The move away from the small molecules therapies model in the post-genomic era. *Creativity and Innovation Management, 20*(4), 311–329.

Suchman, M. C. (1995). Managing legitimacy: Strategic and institutional approaches. *Academy of Management Review, 20*(3), 571–610.

Suddaby, R., & Greenwood, R. (2005). Rhetorical strategies of legitimacy. *Administrative Science Quarterly, 50*(1), 35–67.

Thornton, P. H. (2004). *Markets from culture: Institutional logics and organizational decisions in higher education publishing*. Standford, CA: Stanford University Press.

Thornton, P. H., Ocasio, W., & Lounsbury, M. (2012). *The institutional logics perspective: A new approach to culture, structure, and process*. Oxford: Oxford University Press.

UPS. (2014). *UPS annual report 2013*. Louisville, GA: UPS.

UPS. (2015). *UPS annual report 2014*. Louisville, GA: UPS.

WRI/WBCSD. (2011). *The greenhouse gas protocol: A corporate accounting and reporting standard* (Revised ed.). Washington, DC: World Resources Institute.

Wooten, M., & Hoffman, A. J. (2008). Organizational fields: Past, present and future. In R. Greenwood, C. Oliver, K. Shalin-Andersson, & R. Suddaby (Eds.), *The sage handbook of organizational institutionalism* (pp. 130–147). London: Sage.

World Economic Forum. (2012). *Global agenda councils on logistics & supply chain*. New York, NY: World Economic Forum.

Chapter 7
Synergy between Population Policy, Climate Adaptation and Mitigation

Jane N. O'Sullivan

Abstract Global, national and regional population projections are embedded in projections of future greenhouse gas emissions and in the anticipated impacts of climate change on food and water security. However, few studies acknowledge population growth as a variable affecting outcomes. Neither the uncertainty around population projections nor the scope for interventions to moderate growth is discussed. Instead, a deterministic approach is taken, assuming that population growth is governed by economic and educational advances.

This chapter reviews the treatment of population in climate change scenarios and the prospects for proactive interventions to influence outcomes. Sensitivity analyses have demonstrated population to be a dominant determinant of emissions. The assumption that population growth is determined by economic and educational settings is not well supported in historical evidence. Indeed, economic advance has rarely been sustained where fertility remained above three children per woman. In contrast, population-focused voluntary family planning programmes have achieved rapid fertility decline, even in very poor communities, and enabled more rapid economic advance.

Policy-based projections of global population are presented, based on the historical course of nations that implemented effective voluntary family planning programmes. If remaining high-fertility nations adopted such programmes, global population could yet peak below 9 billion. Current trends make it more likely to exceed 13 billion people by 2100 unless regional population pressures cause catastrophic mortality rates from conflict and famine. Global support for family planning could reduce population by 15% by 2050 and 45% by 2100 compared with the current trend. Co-benefits include gender equity, child health and nutrition, economic advancement, environmental protection and conflict avoidance.

Keywords Population growth • Greenhouse gas emissions • Shared socioeconomic pathways • Economic development • Family planning

---

J.N. O'Sullivan (✉)
School of Agriculture and Food Sciences, University of Queensland,
St Lucia, QLD, 4067, Australia
e-mail: j.osullivan@uq.edu.au

7.1 Introduction

Several recent reports stress that climate change is accelerating and that its impacts may be more severe than earlier models suggested. Hansen et al. (2016, p. 3761) found evidence for acceleration of ice-melt and its relationship with storm intensity and concluded that "The modelling, paleoclimate evidence, and ongoing observations together imply that 2 °C global warming above the preindustrial level could be dangerous". Schleussner et al. (2016) demonstrated considerable difference in impact between 1.5 °C and 2 °C warming, but concluded that at best we may be able to minimise the period during which global mean temperature may temporarily exceed 1.5 °C. They highlighted the gap between current national commitments under the Paris Agreement and the emissions reductions needed to meet that agreement's goal to limit the increase in the global average temperature "to well below 2 °C above pre-industrial levels". Schellnhuber, Rahmstorf, and Winkelmann, 2016 observed that almost no scenarios so far modelled achieve a greater than 50% chance of remaining below 1.5 °C warming. Spratt (2016) emphasised that basing required action on a 50% probability of achieving a "safe" target does not meet any normal standards of risk management. He argued that there is already no carbon budget left if we are to have 90% chance of remaining under 2 °C.

Such urgency emphasises that action to reduce drivers of climate change must be taken on all effective fronts simultaneously. Focusing responses too narrowly will mean other necessary changes are addressed too late. All avenues to reduce emissions should be pursued, unless they compete directly for the same resources. Rockström et al. (2017) suggest a "roadmap" consistent with less than 2 °C warming that requires anthropogenic carbon dioxide emissions to peak before 2020 and halve each decade thereafter.

Yet there is one line of action that has so far been excluded from the discourse of the United Nations Framework Convention on Climate Change (UNFCCC). This is despite it being inexpensive, impacting climate adaptation and mitigation simultaneously, enhancing the impact of all other climate change responses and directly benefiting the poorest and most vulnerable sectors of humanity, particularly women and children in the least-developed countries. This low-hanging fruit is the extension of voluntary family planning and access to birth control, to minimise further growth in the human population.

In models of future greenhouse gas emissions, the contribution of population growth is often buried in model assumptions and uninterrogated in their analysis. For example, in the roadmap of Rockström et al. (2017), renewable energy roll-out is expressed only in terms of percentage share of primary energy. The prospect of a doubling or more in energy demand, and the attribution of this demand growth among population growth, economic development or technology change, is thereby avoided altogether. Such population-energy-technology (PET) analyses have found that the sensitivity of emissions to population change is greater than that to change in GDP per capita by a factor of more than two (Jorgenson & Clark, 2010) to almost seven (Casey & Galor, 2017) when considering only carbon dioxide emissions from fossil fuels and industry (FFI).

Given the less flexible relationship between population and demand for land resources, the common omission of emissions from agriculture, biomass use and land use change underestimates the influence of population on emissions. Bajželj et al. (2014) found that greenhouse gas emissions from the food system were sensitive to population outcomes by a factor of 1.9, meaning that a 10% increase in population would result in 19% more emissions from the food system, assuming the same wealth and dietary preferences.

Using an economic-demographic model, taking account of multiple channels of effects of change in fertility and population on economic growth, Casey and Galor (2017) estimated that moving from the medium to the low variant of the United Nations (2015) global population projection could reduce FFI emissions by 10% by 2050 and 35% by 2100, despite increasing income per capita. O'Neill et al. (2010), including emissions from the food system but using UN population estimates from 2003, estimated emissions reduction around 15% by 2050 and 35–42% by 2100. Their careful analysis accounted for changes in urbanisation, age distribution and household size on a country-by-country basis.

These studies applied alternative population projections as exogenous factors, without identifying to what extent specific measures would achieve lower population outcomes. In this analysis, historical evidence for the effect of both economic advancement and voluntary family planning programmes on population outcomes is examined; future projections are based on assuming the adoption of family planning programmes achieves the average outcome achieved by such programmes in the past. It therefore more directly addresses the value of including population measures in the climate change response.

7.2 Population Assumptions in Climate Change Scenarios

Predicting future greenhouse gas emissions, and the effect of mitigation measures, involves many assumptions about future trends in economic, social and technological change and international relations. To help make such assumptions explicit and consistent between modelling exercises, and to explore the likely range of outcomes, scenario narratives are built describing alternative possible futures.

The socioeconomic scenarios developed by the Intergovernmental Panel on Climate Change (IPCC) have formed the basis of many attempts by independent research groups to model the impact of policy options on outcomes. The "shared socioeconomic pathways" (SSPs) described in the IPCC's Fifth Assessment Report (AR5) (IPCC, 2014) replace the previous SRES scenarios, named from the Special Report on Emissions Scenarios (IPCC 2000). The SSPs are likely to remain the dominant framework for modelling for some years. The SSPs comprise five scenario "families" (O'Neill et al., 2013). In the base case, the scenario describes a future without policies to address climate change, setting assumptions about trends in key drivers and the interactions between them. Against this, modellers may vary policy, technology and other assumptions to generate mitigation scenarios.

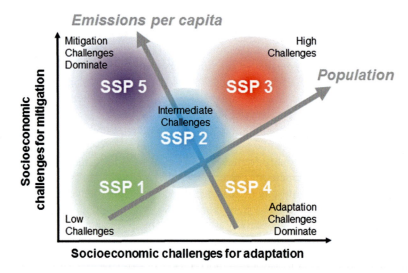

Fig. 7.1 A conceptual map of the five families of IPCC Shared Socioeconomic Pathways (SSPs), in relation to the strength of mitigation and adaptation challenges posed by each scenario (after van Vuuren et al., 2014). Approximate trends in population outcomes and emissions per capita outcomes are superimposed. Population growth is most strongly related to adaptation challenges

Each of the SSPs has a different global population trajectory (van Vuuren et al., 2014). These projections depend primarily on the assumed timing, rate and extent of the fall in family size in remaining high-fertility countries, and to a lesser extent assumptions about change in mortality rates, migration and family size in low-fertility countries (i.e. those below the "replacement rate", around 2.1 children per woman, at which children just replace their parents' generation in the absence of migration). However, the SSPs do not differ with respect to actions to influence fertility—none are included in any of the scenarios. This is because family size outcomes are assumed to be a product of economic and educational outcomes (Samir & Lutz, 2014).

Figure 7.1 indicates the relative challenges posed for climate change adaptation and mitigation by each SSP baseline scenario and the approximate trends in population and emissions per person among the SSPs (their actual population projections are given in Fig. 7.7).

Notably, all but one (the worst-case scenario, SSP3) of the SSP scenario families anticipate a global population well below the UN's current medium projection—indeed, below the 95% probability range of the UN's 2015 probabilistic projections (UNDESA, 2015). The preferred scenario, SSP1, combines a very low global population with low per capita footprint in a world of more integrated and equitable governance. SSP5 combines a similarly low population path with high energy demands per person. For these two scenarios, the population path is lower than the UN's "low-fertility projection", which is not intended as a realistic scenario but as a sensitivity analysis: it merely applies a fertility rate (the average number of children per woman) 0.5 units lower than the medium projection in all countries with immediate effect. Fertility is expected to fall steadily under the UN's medium projection, but no faster or slower in the low and high projection after the initial adjustment of 0.5 units. However, a path similar to the SSP1 and UN low projection is achievable if fertility were reduced in remaining high-fertility countries faster and further than assumed in the UN projections.

The population outcome has enormous impact on prospects for both mitigation and adaptation, as Young, Mogelgaard, and Hardee (2009) also demonstrated with respect to the earlier SRES scenarios. Riahi, van Vuuren, and Kriegler (2017) reported that, across six independent integrated assessment models running a total of 105 mitigation scenarios, outcomes as low as 2.6 W/m^2 climate forcing (consistent with less than 2 °C warming) were found to be infeasible when applying SSP3. This was despite SSP3 assuming considerably less economic development than other scenarios. SSP3 also had no feasibility of increasing forest cover, and only SSP1 projected forest expansion as likely in the base scenario.

It is vital to note that each of the UN's revisions in the past decade has increased the expected population, because fertility decline is not happening as fast as its medium scenario expects. The UN's estimate for the year 2100 has increased by more than two billion in just 11 years (Fig. 7.2). This suggests that the feasibility of more favourable climate change outcomes is being eroded over time, as global population growth is exceeding the expectations of lower-emissions models.

The reason for these regular upward revisions is obvious when we look at annual increments of global population growth (Fig. 7.3). In many countries, rapid falls in fertility were occurring from the 1970s to 1990s, which enabled the global population increment to peak in 1988 and to decline throughout the 1990s. However, since 2000, the increment has increased again. This is the result of fertility decline slowing, stalling or reversing since the withdrawal of funding and political support for family planning programmes from the mid-1990s (Bongaarts, 2008). Countries such as Indonesia, Algeria and Egypt, which achieved considerable fertility decline under family planning programmes prior to the mid-1990s, have seen fertility increase again before reaching replacement rate. This reversal occurred despite accelerated progress on girls' education, child mortality and poverty reduction (factors popularly claimed to drive fertility decline), as these were high priority targets of the Millennium Development Goals (MDGs). The dramatic fall in international support for family planning (Fig. 7.3, right-hand axis) was the factor most consistent with this reversal (Sinding, 2009), providing evidence that its influence on population growth has been stronger than is commonly recognised. The goal of achieving universal access to sexual and reproductive health services was belatedly added to the MDGs in 2007, but remained the least addressed in its agenda.

Interruptions or reversals of the fertility transition, such as those in Egypt and Indonesia, are not accommodated in the UN's population model (O'Sullivan, 2016). The UN's medium projection continues to expect the downward trend in global increment to resume, based on immediate resumption of fertility decline in all high-fertility countries, despite many of them seeing little if any decline recently. As shown in Fig. 7.3, data on actual change in global population, reported annually

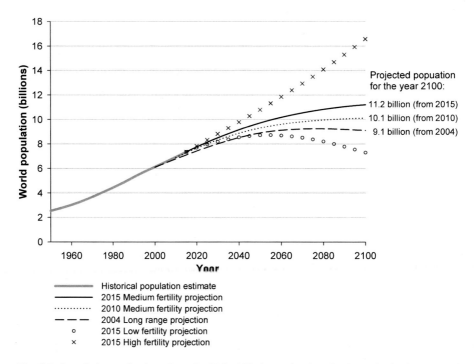

Fig. 7.2 Population projections from the United Nations, showing the dramatic rise in expected outcomes since 2004. (Data sources: UNDESA, 2004, 2011, 2015)

by the Population Reference Bureau (PRB, 2011–2016), are greatly exceeding the "medium" projection and instead are approximating the "constant fertility" projection, which assumes all countries continue with the same fertility they had in 2010. If sustained, it results in a global population of 28 billion by 2100. Of course, such a population could not be fed—if fertility does not fall, then death rates must rise.

Yet, despite the poor track record of recent projections, most climate impact modellers continue to regard future population as predetermined. They do not consider what measures might be available to influence it. Most do not even acknowledge the uncertainty of population projections or consider any sensitivity analysis to see how emissions outcomes would be affected if population is higher or lower than expected.

Marangoni et al. (2017) attempted sensitivity analyses on the main drivers of FFI emissions in the SSP scenarios and concluded population growth was relatively unimportant compared with economic growth and energy intensity of the economy, although the latter two tended to offset each other. This outcome was a product of their methodology, which contrasted outcomes to 2050 in SSP2 projections when individual drivers were substituted from SSP1 or SSP3. Hence it highlighted whichever factors varied most in the short term between these three SSP scenarios, due to the arbitrary assumptions underlying those scenarios. They failed to acknowledge

7 Synergy between Population Policy, Climate Adaptation and Mitigation

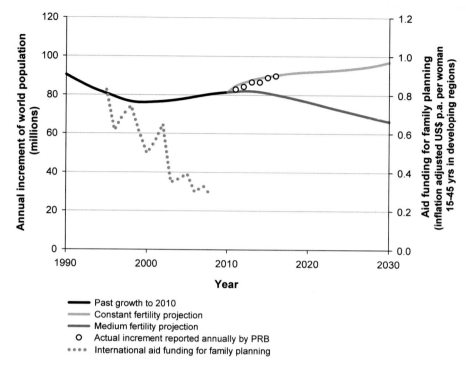

Fig. 7.3 The annual increment of global population 1990–2010, and that projected under the UN's medium fertility and constant fertility projections (UNDESA, 2013). Black circles give estimates of actual increment reported annually in the Population Reference Bureau's "World Population Datasheets" (PRB, 2011–2016). International aid spending on family planning is plotted against the right axis. (Data source: UN Economic and Social Council, 2010)

that energy intensity of the economy and growth in GDP per capita are not independent factors but tend to offset each other—increasingly so as energy constraints intensify in the future. In most cases, if a higher assumption is imposed about future economic growth, a lower energy intensity is required to achieve it, so that the net effect is predictably smaller than the individual effects. The effect of population was further understated by omitting land use emissions and biomass sequestration.

In contrast, sensitivity analyses based on historical data have found population growth to be a stronger driver of emissions than economic growth. The studies of Jorgenson and Clark (2010), Casey and Galor (2017), O'Neill et al. (2010) and Bajželj et al. (2014) have already been mentioned. Alexander et al. (2015) found that population growth has been the largest driver of land use change, although dietary changes in emerging economies are an increasingly important contributor. The World Resources Institute's exemplary series Creating a Sustainable Food Future found that achieving replacement level fertility in sub-Saharan Africa by 2050 would spare an area of forest and savannah larger than Germany from conversion to cropland, and in doing so save 16 Gt of carbon dioxide emissions (Searchinger et al., 2013). The RoSE project, a major international effort to model energy and emissions pathways, discovered that a higher-than-expected population had a far greater impact on deforestation and land use-related emissions than high economic growth (Kriegler, Mouratiadou, Luderer, et al., 2013).

As illustrated in analyses such as those by O'Neill et al. (2010) and Casey and Galor (2017), the impact of a modelled change in demographic drivers is relatively small until mid-century, but expands greatly in the second half of this century. The slow divergence of population outcomes after policy change ("demographic momentum") is often used to argue that population measures should have lower priority compared with energy sector interventions. It should instead be a reason for even greater urgency, to ensure that the substantial gains are not further deferred, and that a higher peak population does not render safe climate scenarios infeasible. Neglect of population policy over the past two decades has likely added more than two billion people to the global peak population (Fig. 7.2). The low priority afforded to population measures implies that they would compete with other mitigation or adaptation efforts. This is a mistake, since family planning measures are usually cost-negative, saving health and education systems far more than they cost. Hence, they simultaneously liberate resources for other climate change and development measures and reduce the scale of other measures required to meet humanity's needs.

7.3 Relationship between Enrichment and Fertility Decline

Lower population outcomes may be widely recognised as beneficial for climate change adaptation and mitigation, and a faster decline in fertility in developing countries is accepted as necessary to achieve a lower population. However, there is disagreement regarding whether direct interventions are effective and appropriate to speed the fertility decline. The SSP scenarios assume that the low population outcomes can be achieved as a result of indirect effects of economic development and education, without any interventions directly aimed at lowering fertility. The historical record does not provide strong evidence for this position. No country has been able to achieve significant enrichment while fertility and population growth remained high, with the exception of those with large mineral resources. The latter, including Syria and Egypt, did not see fertility falling rapidly as a result of oil wealth, and have suffered a reversal of fortune as increasing dependence on food imports coincides with declining oil revenue (Ahmed, 2017).

In contrast, there is abundant evidence that population-focused voluntary family planning programmes were highly effective in causing rapid fertility decline and subsequently accelerated economic development. Countries such as South Korea, Thailand and Costa Rica, in which voluntary family planning was extended and promoted even to poor, rural and remote communities, saw rapid fertility decline, two to three times as fast as UN projections expect for remaining high-fertility countries (Fig. 7.4). They subsequently experienced broad-based economic development, accelerating only after fertility fell below three children per woman and population growth slowed. The timing of their fertility transition matched that of

7 Synergy between Population Policy, Climate Adaptation and Mitigation

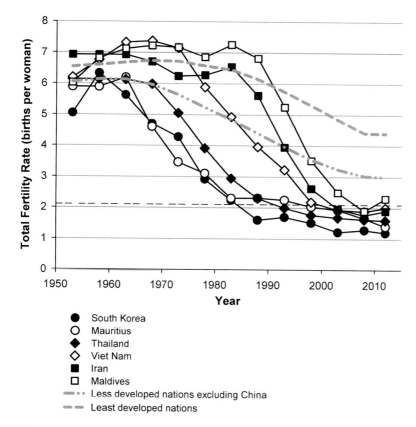

Fig. 7.4 Time course of total fertility rate (TFR, births per woman) for selected countries that implemented population-focused voluntary family planning programmes at differing times, showing rapid change in fertility, compared with aggregate TFR for less developed countries (excluding China) and least-developed countries. The horizontal dashed line approximates "replacement level" fertility. (Data from UNDESA, 2015)

their family planning programmes, with no apparent economic or educational trigger. Meanwhile, some countries whose wealth and female education levels were above regional averages, such as the Philippines, Malaysia and Nigeria, saw little fertility decline. In several countries where family planning programmes were neglected before reaching replacement rate, such as Indonesia, Bangladesh, Algeria and Ghana, the fertility decline stalled and in some cases reversed.

Notwithstanding that these family planning-driven fertility transitions generally preceded economic and educational improvement, the overall correlation between GDP per person and total fertility rate of nations has led to the common assumption that economic development drives the fertility transition. To further investigate the direction of causation, Fig. 7.5 explores whether the level of wealth affected fertility decline, or whether fertility affected economic development. For all countries in each 5-year interval for which data were available, the change in fertility over 5 years was plotted against the level of GDP per capita at the start of the period

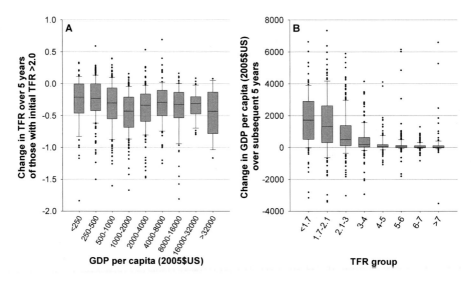

Fig. 7.5 (**a**) The rate of fertility decline as a function of level of wealth and (**b**) rate of economic development as a function of level of fertility. Data points represent each country in each 5-year period between 1960 and 2010. All countries and time periods with available data are included. Box plots span 25th percentile, median and 75th percentile and whiskers extend to the 10th and 90th percentile. GDP per capita (inflation-adjusted US$2005) are from the World Bank economic database and fertility data from UNDESA (2015)

(Fig. 7.5a). It was found that the rate of fertility decline was unrelated to per capita GDP. The poorest countries could reduce fertility as rapidly as middle-income countries if they were motivated to do so. Conversely, when the change in GDP per capita was plotted against the total fertility rate (TFR) prevailing at the start of each interval, it is evident that economic development has been severely hampered by high fertility (Fig. 7.5b). When fertility remains above three children per woman, the chance of sustained economic improvement has proven to be extremely low. While low fertility has not guaranteed enrichment in any 5-year period, over 20-year intervals all low-fertility countries made considerable gains in wealth, including those with shrinking populations. Fertility decline appears to be a necessary, if not sufficient, precondition for economic development.

Figure 7.6 contrasts the experience of all countries that had high fertility in 1950, grouped according to their maximum rate of fertility decline over any 20-year period. Group 1 contained countries that had TFR above 5 at the start of the series (1950–1955) and where TFR fell after a particular date, at a rate exceeding 1.5 units per decade, to near or below replacement (unless insufficient time had elapsed since the start of decline). It was verified that each of these countries adopted voluntary family planning measures around the time that the birth rate began to fall, although not all maintained them until reaching replacement level fertility. Group 2 also showed considerable decline in TFR after a given date, but at a slower rate, between 0.5 and 1.5 units per decade. Group 3 showed no distinct start date for fertility decline and still have fertility rates above 4. Group 4 (not plotted) were considered "post-transition" countries that had fertility rates below 5 and falling at the start of

7 Synergy between Population Policy, Climate Adaptation and Mitigation   113

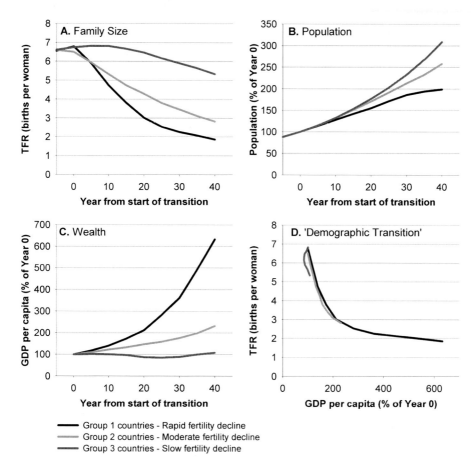

Fig. 7.6 The averaged time course for (**a**) fertility, (**b**) population and (**c**) GDP per capita (inflation-adjusted US$), and (**d**) the relationship between total fertility rate (TFR) and per capita GDP, for developing countries grouped according to the rate of their fertility transition. Each of the rapid transition countries (Group 1) deployed successful family planning programmes. Many countries in Group 2 had programmes that were weaker or not sustained. Group 3 countries generally did not have widespread family planning efforts. Year 0 is the start of the fertility transition in each country, or 1970 for weak adopters (Group 3). Population and fertility data from UNDESA 2013, economic data from World Bank economic database

the data series. Countries where either immigration or emigration contributed more than 15% to population change were excluded. Before averaging each group's data, they were synchronised with respect to the timing of their fertility transition by designating Year 0 to be the start of the transition, or 1970 where no distinct change in fertility path is observed (Group 3).

The sixteen countries included in Group 1 were Algeria, Bangladesh, Bhutan, Cambodia, Chile, China, Costa Rica, Iran, South Korea, Libya, Maldives, Mongolia, Oman, Thailand, Tunisia and Viet Nam. Group 1 countries rejected due to high migration were Hong Kong, Macao, Singapore, Aruba, Kuwait and Saudi Arabia (high immigration), and Mauritius and Guyana (high emigration). Note that China is included despite the presence of coercive programmes from 1979 because most of the fertility decline occurred under the voluntary family planning programme in place from 1970 to 1978. Note also that the data are not population-weighted: China's data have no more influence on the average than those of Bhutan or Maldives. Group 2 contained 39 included countries and 48 high-migration countries, group 3 contained 26 included countries and 19 high-emigration countries.

Only Group 1 countries have achieved a tapering of their population growth (Fig. 7.6b). Most of these will achieve a peak population around 2–2.5 times the population when they started addressing family planning. Group 2 countries have lessened population growth but have not reduced family size as fast as the number of families has increased. Hence, most are still adding more people each year than ever before. Group 3 countries have seen their population triple in the same period. Due to their high proportion of young people yet to start families, they have another doubling in store if they choose to embrace family planning now (and more if they do not).

The impact of fertility decline on wealth can be seen dramatically in Fig. 7.6c. Rapid fertility decline has been associated with dramatic economic improvement. Slow-transition countries have seen virtually none. Figure 7.6d plots the fertility rate as a function of wealth. It contains two features that are at odds with the popular belief that development drives fertility decline:

1. The relationship between fertility and wealth is steeply concave, as *fertility fell first* before economic development accelerated.
2. *All three groups have followed the same path.* Those that accelerated fertility decline progressed more rapidly to economic development. With rare exceptions, "development first" has not been an achievable option.

These results do not suggest that ending poverty does not influence family size, but that reducing poverty as a population control strategy simply has not proven possible for most high-fertility countries. High population growth poses such a high economic burden, together with ongoing crowding of limited natural resources, that significant reductions in poverty have been impossible to achieve. Family planning programmes, on the other hand, have proven achievable even in the poorest settings, and the economic benefit to families has translated into societal enrichment.

7.4 How Much could Population Policy Measures Contribute to the Climate Change Response

Although the effects of a lower population on greenhouse gas emissions have been previously estimated (e.g. Casey & Galor, 2017; O'Neill et al., 2010), the effects of policies and programmes intended to reduce population have not been quantified. To address this question, population projections were compared under two scenarios: "business as usual" and a proactive family planning scenario. In the latter, all countries that currently have above-replacement fertility adopt voluntary, client-focused and culturally appropriate family planning measures, which seek to give all women the means to avoid unwanted pregnancy and which address traditional barriers to fertility regulation. It was assumed that each country would achieve the average path of fertility decline achieved in the past by Group 1 (as depicted in Fig. 7.6a), starting from their 2010–2015 fertility rate. This is a conservative scenario, as better contraceptive technologies, communications, education levels and community engagement methods all have the potential to make future programmes more effective than in the past. However, the projection is also quickly outdated: each year of delay in implementing such programmes would add around 100 million people to the achievable global peak population.

The business-as-usual scenario would see the international community continue to direct a derisory level of funding and programme attention to family planning (which receives less than 1% of international aid) and family planning programmes continue to lack the scale and visibility needed to reach the majority of disadvantaged people and to achieve rapid fertility decline. The fertility path applied to high-fertility countries (Group 3, Fig. 7.6a) continued forward using the UN's medium projection. However, each country's starting point on this path matched their current fertility, so that countries with current fertility above 5.5 experienced a further period of slow change before reaching the more rapid phase modelled in the UN projection. In addition, it was assumed that very low-fertility countries (below 1.5) were successful in boosting birth rates and achieved the UN's high projection (while remaining below replacement rate fertility).

Under these assumptions, a business-as-usual approach is likely to see growth exceed the UN's current medium projection, reaching 13 billion before the century's end (Fig. 7.7). It is questionable whether such a population would be achieved, but the risk is that it will be curtailed by widespread conflict and famine. Apart from being a catastrophe in its own right, such a scenario would likely derail climate action.

On the other hand, if the remaining high-fertility countries were to embrace voluntary family planning programmes, a global population path close to the UN's current low projection could be achieved. This would put the IPCC's best-case, SSP1 scenarios into the realm of possibility.

Although the proactive family planning scenario achieves a similar population outcome to the SSP1 model and the UN low-fertility variant, it does so in a very different and more plausible way. It is important to recognise that an avoided birth has more impact on future population if it occurs sooner rather than later. Delayed fertility decline allows the demographic momentum to build, so that greater fertility decline is ultimately required to achieve the same population. The UN's low-fertility projection is unrealistic, because while it assumes a very modest reduction in fertility (half a child per woman) compared with the medium scenario, it applies this reduction within the first 5-year interval. The SSP1 model, on the other hand, relies on future improvements in education and economic conditions to drive fertility decline, resulting in a later and more gradual decline, but declining to an extraordinary extent, settling around 1.2 children per woman across today's developing

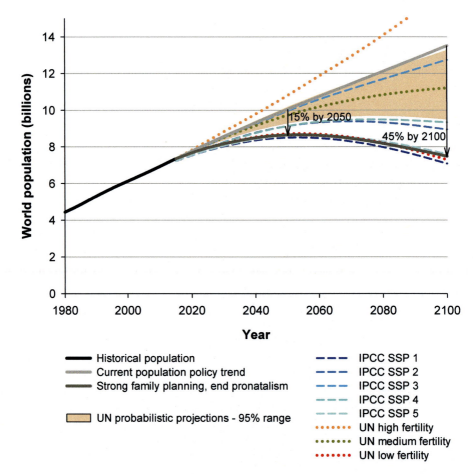

Fig. 7.7 Projections of future global population based on population policy, comparing a business-as-usual scenario (countries continue their recent trends) and a proactive scenario (remaining high-fertility countries adopt nationwide voluntary family planning programmes, achieving the average fertility path that past programmes achieved, and low-fertility countries abandon attempts to increase births). Also depicted for comparison are the UN high-, medium- and low-fertility projections and 95 % probability range (UNDESA 2015) and the IPCC's Shared Socioeconomic Pathways (Samir & Lutz, 2014)

countries, despite developed countries remaining or rebounding to around 1.7 children per woman (Wittgenstein Centre for Demography and Global Human Capital, 2015). No explanation is offered for this astonishing assumption. In the proactive family planning model offered in the current study, fertility decline is rapid over the first three decades as a result of family planning programmes, but is expected to stabilise fertility not far below replacement, around 1.8–2.0 children per woman.

This analysis concludes that without urgent promotion of voluntary family planning, the population projection of SSP1 is not possible and "safe" climate change pathways are unlikely. In a world of 11 billion people, let alone 14 billion, considering the requirements for arable land, fresh water, nitrogenous fertiliser and building materials, even assuming 100% renewable energy, net drawdown of atmospheric carbon would require implausible levels of carbon capture and storage. It seems rash to turn our backs on technologies as cheap, reliable and widely beneficial as voluntary family planning in favour of technologies as unproven, costly and potentially damaging as carbon capture and storage.

7.5 Empowering Women and Couples to Control their Fertility is the Most Cost-Effective Climate Action

Wheeler and Hammer (2010) argued that climate finance could provide a much-needed boost to the long-neglected goal of universal reproductive rights. Using country-by-country analyses, they found that avoiding unwanted births through investments in family planning and girls' education would avoid greenhouse gas emissions at considerably lower cost than renewable energy initiatives, and lower than most reforestation initiatives. This was true even in countries where per capita emissions are very low. In more than 60 countries, the cost was less than $10 per tonne. Although spending on family planning avoided more births than spending on girls' education, the synergy between these activities meant that lowest-cost emissions reductions were achieved with a combination of both interventions. While the emissions reduction alone would justify the allocation of funds, the same investment would simultaneously empower women and improve health, nutrition, education and economic outcomes for families.

A USAID study of 16 sub-Saharan African countries in 2006 found that fulfilling the unmet need for family planning would not only contribute materially to the attainment of all other MDGs, but each dollar spent on family planning saved between two and six dollars by reducing the need for interventions to meet other development goals (Moreland & Talbird, 2006).

There is a high rate of unwanted pregnancy even in developed countries, where each avoided birth reduces far more emissions. Even in developed country contexts, net costs have been found to be negative. A recent programme to reduce teen pregnancies in the US state of Colorado lowered the teen birth rate and abortion rate by 40% and 42%, respectively, and saw a similar decline in the number of unintended pregnancies in unmarried women under the age of 25 (Tavernise, 2015). The programme saved Medicaid around $5.85 in perinatal care for every $1 invested (Colorado Department of Public Health and Environment, 2015).

It must be stressed that the successful voluntary family planning programmes of the past did not rely solely on ensuring access to contraception. Large desired family size remains the main determinant of high fertility. Even among those women who do not want to become pregnant, social and spousal pressure and misconceptions about side effects are more commonly cited reasons for not using modern contraception than lack of access and affordability (Ryerson, 2010). The successful programmes promoted the benefits of fewer, more widely spaced children, employed culturally appropriate means to change social norms around family size and women's roles, and addressed the many barriers to achieving fertility regulation (Campbell, Sahin-Hodoglugil, & Potts, 2009).

More recently, Population Health and Environment programmes, which integrate family planning with livelihood, public health and environmental management interventions, are showing that coherent cross-sectoral programmes can greatly increase community acceptance of (and even enthusiasm for) family planning, overcoming cultural resistance (PAI et al., 2015). Even in developed countries, where unintended pregnancy is associated with negative outcomes for women and children (Brookings Institute, 2011), proactive advice on fertility regulation is proving effective (Oregon Foundation for Reproductive Health, 2012).

All these programmes are consistent with the UN Programme of Action adopted at the 1994 International Conference on Population and Development (UNFPA, 1994), widely referred to as the Cairo Agenda, which is still upheld as the current international treaty on population. The Cairo Agenda stresses the negative impacts of population growth on the environment and on the alleviation of poverty and advocates responsible parenthood, in which parents should "take into account the needs of their living and future children and their responsibilities towards the community" (Para 7.3). The extent to which population growth heightens risks of climate change impacts should be among the considerations of prospective parents. Yet such a discourse is shunned by the UN and the development community in the name of the Cairo Agenda, wrongly claiming that demographic agenda by their nature conflict with reproductive rights (Campbell, 2014). This is a misrepresentation of the Cairo Agenda, which states (Para 7.12) that "Demographic goals, while legitimately the subject of government development strategies, should not be imposed on family planning providers in the form of targets or quotas for the recruitment of clients". Many successful family planning programmes prior to 1994 have demonstrated the strong synergy between pursuing demographic agenda and elevating women's health and rights. The post-1994 taboo on demographic goals has had tragic consequences, not least for women's reproductive health and rights, but also for the security of the next generation in the face of climate change.

7.6 High-Fertility Countries Face Multiple Challenges

As many commentators have observed, the effects of population change in sub-Saharan Africa dwarf the likely impacts of climate change on food and water security and on environmental damage. Fresh water access is a critical determinant of community resilience in this regard (Vörösmarty, Green, Salisbury, & Lammers, 2000). It has been estimated that we will need additional fresh water equivalent to 20 Nile Rivers to feed one billion more people (Bigas, 2012). Currently we are adding one billion every 12 years or less. Zeng, Neelin, Lau, and Tucker (1999) found strong evidence that the reduction in rainfall in West Africa over recent decades was due in larger part to regional vegetation change (deforestation) than to global climate change. Carter and Parker (2009, p. 676) evaluated threats to groundwater access in Africa, and observed:

> The climate change impacts [on groundwater] are likely to be significant, though uncertain in direction and magnitude, while the direct and indirect impacts of demographic change on both water resources and water demand are not only known with far greater certainty, but are also likely to be much larger. The combined effects of urban population growth, rising food demands and energy costs, and consequent demand for fresh water represent real cause for alarm, and these dwarf the likely impacts of climate change on groundwater resources, at least over the first half of the 21st century.

Figure 7.8 demonstrates how dramatically the projected increase in population will affect African countries' ability to feed their own populations in comparison with the modest changes in rainfall anticipated by Carter and Parker (2009). Food import dependence is growing in many African and Middle Eastern countries already (Worldwatch Institute, 2015), exposing them to global food price spikes that have been shown to be powerful triggers of civil unrest and violent conflict (Lagi, Bertrand, & Bar-Yam, 2011). The dashed lines in Fig. 7.8 demonstrate how much this challenge could be alleviated if these countries emulated the voluntary family planning successes of the past (using the same model described for Fig. 7.7).

While many commentators acknowledge the future risk of violence and displacement caused by population pressure, almost none are willing to recognise its role in current crises. The Population Institute's's (2015) Demographic Vulnerability Report noted:

> Population pressures are also contributing to environmental degradation and political instability. In effect, rapid population growth is a challenge multiplier, and for many developing countries the challenges are formidable.

A UK all-party parliamentary committee on population and development also warned (APPG, 2015):

> Population dynamics interact with climate change and with conflict to affect people and communities, and will increasingly do so over the course of the 21st century. If the world is to achieve sustainable development then there is an urgent need to scale up access to family planning, and to support sexual and reproductive health and rights.

Moreland and Smith (2012) found that even a modest increase in the rate of fertility decline in Ethiopia would negate the anticipated impacts of climate change on food security. Thankfully, Ethiopia and Rwanda are now making strong progress to extend and promote family planning, but most other east African countries are doing less well.

Apart from the density of population limiting access to natural resources and their services (particularly for the provision of food), the rate of population growth itself presents a formidable economic challenge to high-fertility countries. O'Sullivan (2012) demonstrated that the cost of providing infrastructure, equipment and trained service providers to cater for population growth greatly exceeds the extra economic activity that the additional people can generate. Using long time-series of actual national expenditure in the UK, O'Sullivan (2013) found that it takes around 7% of GDP to add 1% to the capacity of the nation's infrastructure in order

120   J.N. O'Sullivan

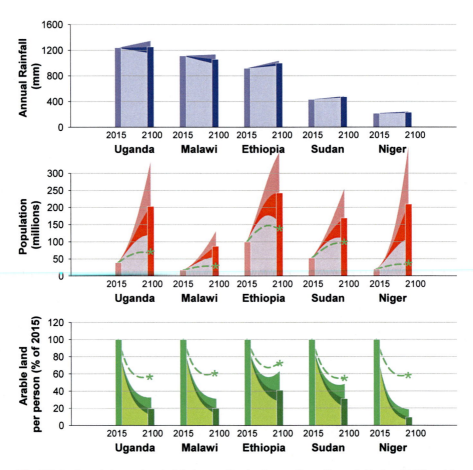

Fig. 7.8 Projected change in rainfall due to climate change (from Carter & Parker, 2009) and in population (from UNDESA, 2015) in five sub-Saharan African countries. Histograms give the medium projection, and area plots indicate the likely range (for population, this is the 80% range using UN probabilistic projections; UNDESA, 2015). Arable land per person is calculated as a percentage of that currently available. The dashed pathways are those that would be achievable if the proactive scenario described in Fig. 7.7 were rapidly initiated

to accommodate 1% more people. An African country growing at 2.5% per annum might require 17% or more of its GDP to be diverted to infrastructure creation that achieves no improvement, merely running in order to stand still against the tide of population growth. This impost, rather than any shortage of land or water, explains the near-absence of economic development in high-fertility countries presented in Figs. 7.5 and 7.6. Its alleviation is likely to have played a greater role in the "Asian Tiger" economies than the more commonly cited "demographic dividend" resulting from shifts in population age distribution (Bloom & Williamson, 1998).

Mutunga and Hardee (2010) reviewed the National Adaptation Plans for Action prepared by least-developed countries during the UNFCCC's 2009 climate adaptation agenda. They found that 37 out of 41 NAPAs highlighted population growth and density as factors increasing vulnerability, but only one proposed project included a population component and none were funded. The UNFCCC guidelines lacked appropriate categories in which population action could be presented as valid climate action. Without including discussion of population dynamics in the climate change discourse, this situation is unlikely to be remedied.

7.7 Conclusion

A safe climate future depends on minimising further growth in the human population. Strengthening efforts to empower women, to avoid unwanted births and to popularise smaller families through voluntary, rights-based family planning programmes are necessary measures without which low-emissions scenarios cannot be achieved (Guillebaud, 2016).

The IPCC socioeconomic scenarios that achieve low emissions, consistent with less than 2 °C global warming, assume very low population growth, far lower than current UN expectations (Riahi et al., 2017; Schellnhuber et al., 2016). Yet no programmes are included in these scenarios to achieve this lower population because it is assumed that strong economic development and educational advances will achieve it. Evidence presented in this chapter finds that:

(a) Current trends are for a higher, not lower, population than current UN expectations, rendering most future emissions scenarios invalid.
(b) In most countries, economic advance has not been a major driver of fertility decline; on the contrary, fertility decline, driven by voluntary family planning programmes, has enabled economic advance. Such programmes have been neglected in recent decades, to the great detriment of the world's poorest people and their environment.

Restoring the international support for voluntary family planning programmes, which existed in the 1970s and 1980s, could reduce the peak human population by many billions. Such action would reduce emissions at lower cost than almost all other options, while simultaneously improving climate change resilience of disadvantaged communities and achieving a wide range of co-benefits with respect to health, the status of women, economic development of least-developed nations, nutrition and food security, conflict avoidance and protection of biodiversity.

The United Nations' Sustainable Development Goals include target 3.7: "By 2030, ensure universal access to sexual and reproductive healthcare services, including for family planning". Few people appreciate that most other Sustainable Development Goals targets depend on achieving this long-neglected goal (Starbird, Norton, & Marcus, 2016).

As Cleland et al. (2006) emphasised, "No contradiction needs to exist between respect for reproductive rights and strong advocacy for smaller families and for mass adoption of effective contraceptive methods". Past population-focused family planning programmes greatly accelerated the empowerment of women and the adoption of more liberal social attitudes to women's roles and rights. Their daughters benefited from the improvement in social attitudes, from less economic strain in the household and from greater parental investment in their outcomes, including better nutrition and improved access to schooling and employment opportunities. The economic and environmental benefits for entire nations were evidently substantial. All of these improvements lessen vulnerability to the impacts of climate change. Yet these manifold benefits have not proven compelling enough for the international community to provide the modest resources needed to extend reproductive freedom to women in the remaining high-fertility countries. The imperative of avoiding dangerous climate change may be the incentive that is needed to harvest what is clearly the low-hanging fruit for sustainable development.

References

Ahmed, N. M. (2017). *Failing states, collapsing systems: biophysical triggers of political violence.* Springer. Retrieved from http://www.springer.com/us/book/9783319478142
Alexander, P., Rounsevell, M. D. A., Dislich, C., Dodson, J. R., Engström, K., & Moran, D. (2015). Drivers for global agricultural land use change: The nexus of diet, population, yield and bioenergy. *Global Environmental Change, 35*, 138–147.
APPG. (2015). *Population dynamics and the sustainable development goals.* A report by the UK All-Party Parliamentary Group on Population, Development and Reproductive Health, July 2015. Retrieved from http://www.appg-popdevrh.org.uk/Population%20Dynamics%20and%20the%20Sustainable%20Development%20Goals.pdf
Bajželj, B., Richards, K. S., Allwood, J. M., Smith, P., Dennis, J. S., Curmi, E., & Gilligan, C. A. (2014). Importance of food-demand management for climate mitigation. *Nature Climate Change, 4*, 924–929. Retrieved from http://www.nature.com/nclimate/journal/v4/n10/full/nclimate2353.html.
Bigas, H. (Ed.). (2012). *The global water crisis: Addressing an urgent security issue. Papers for the InterAction Council, 2011–2012.* Hamilton, Canada: UNU-INWEH. Retrieved from http://inweh.unu.edu/wp-content/uploads/2013/05/WaterSecurity_The-Global-Water-Crisis.pdf.
Bloom, D. E., & Williamson, J. G. (1998). Demographic transitions and economic miracles in emerging Asia. *The World Bank Economic Review, 12*, 419–455.
Bongaarts, J. (2008). Fertility transitions in developing countries: Progress or stagnation? *Studies in Family Planning, 39*(2), 105–110.
Brookings Institute. (2011). *The high cost of unintended pregnancy* (Report). Retrieved from https://www.brookings.edu/research/the-high-cost-of-unintended-pregnancy/
Campbell, M. (2014). Ending the silence on population. In M. Hempel (Ed.), Facing the population challenge: Wisdom from the elders (pp. 65–71). Redlands, CA: Blue Planet United. Retrieved July 23, 2017 from https://www.facebook.com/pg/BluePlanetUnited/about/?ref=page_internal.
Campbell, M., Sahin-Hodoglugil, N. N., & Potts, M. (2009). Barriers to fertility regulation: A review of the literature. *Studies in Family Planning, 37*(2), 87–98.
Carter, R. C., & Parker, A. (2009). Climate change, population trends and groundwater in Africa. *Hydrological Sciences Journal, 54*(4), 676–689.
Casey, G., & Galor, O. (2017). Is faster economic growth compatible with reductions in carbon emissions? The role of diminished population growth. *Environmental Research Letters, 12*, 014003. https://doi.org/10.1088/1748-9326/12/1/014003.

Cleland, J., Bernstein, S., Ezeh, A., Faundes, A., Glasier, A., & Innis, J. (2006). Family planning: The unfinished agenda. *Lancet, 368*(9549), 1810–1827.

Colorado Department of Public Health & Environment. (2015). *Preventing unintended pregnancies is a smart investment.* Retrieved from https://web.archive.org/web/20151110092403/https://www.colorado.gov/pacific/sites/default/files/HPF_FP_UP-Cost-Avoidance-and-Medicaid.pdf

Guillebaud, J. (2016). Voluntary family planning to minimise and mitigate climate change. *British Medical Journal, 353*, i2102. https://doi.org/10.1136/bmj.i2102.

Hansen, J., Sato, M., Hearty, P., Ruedy, R., Kelley, M., Masson-Delmotte, V., et al. (2016). Ice melt, sea level rise and superstorms: Evidence from paleoclimate data, climate modeling, and modern observations that 2 °C global warming could be dangerous. *Atmospheric Chemistry and Physics, 16*(6), 3761–3812. https://doi.org/10.5194/acp-16-3761-2016.

IPCC (2000). *Special Report on Emissions Scenarios.* A Special Report of Working Group III of the Intergovernmental Panel on Climate Change. Cambridge, UK: Cambridge University Press.

IPCC. (2014). In O. Edenhoger (Ed.), *Climate change 2014: Mitigation of climate change: Working group III contribution to the fifth assessment report of the intergovernmental panel on climate change.* Cambridge: Cambridge University Press.

Jorgenson, A. D., & Clark, B. (2010). Assessing the temporal stability of the population/environment relationship in comparative perspective: A cross-national panel study of carbon dioxide emissions, 1960–2005. *Population and Environment, 32*, 27–41.

Kriegler, E., Mouratiadou, I., Luderer, G., et al. (2013). *Roadmaps towards sustainable energy futures and climate protection: A synthesis of results from the RoSE project* (1st ed.). Potsdam: Potsdam Institute for Climate Impact Research.

Lagi, M., Bertrand, K. Z., Bar-Yam Y (2011). *The food crises and political instability in North Africa and the Middle East.* Cambridge, MA: New England Complex Systems Institute. Retrieved from http://arxiv.org/pdf/1108.2455.pdf

Marangoni, G., Tavoni, M., Bosetti, V., Borgonovo, E., Capros, P., Fricko, O., ... van Vuuren, D. P. (2017). Sensitivity of projected long-term CO_2 emissions across the shared socioeconomic pathways. Nature Climate Change 7:113–117. https://doi.org/10.1038/nclimate3199., http://www.nature.com/nclimate/journal/v7/n2/full/nclimate3199.html

Moreland, S., & Smith, E. (2012). *Modeling climate change, food security and population: Pilot testing the model in Ethiopia.* Washington, DC: Futures Group with MEASURE Evaluation PRH. Retrieved from https://www.measureevaluation.org/resources/publications/sr-12-69.

Moreland, S., & Talbird, S. (2006). *Achieving the Millennium Development Goals: The contribution of fulfilling the unmet need for family planning.* Washington, DC: USAID. Retrieved from http://pdf.usaid.gov/pdf_docs/Pnadm175.pdf.

Mutunga, C., & Hardee, K. (2010). Population and reproductive health in national adaptation programmes of action (NAPAs) for climate change. *African Journal of Reproductive Health, 14*(4), 133–146.

O'Neill, B. C., Dalton, M., Fuchs, R., Jiang, L., Pachaui, S., & Zigova, K. (2010). Global demographic trends and future carbon emissions. *Proceedings of the National Academy of Sciences of the United States of America, 107*, 17521–17526.

O'Neill, B. C., Kriegler, E., Riahi, K., Ebi, K., Hallegatte, S., Carter, T. R., et al. (2013). A new scenario framework for climate change research: The concept of shared socio-economic pathways. *Climate Change Special Issue, 122*(3), 387–400. https://doi.org/10.1007/s10584-013-0905-2.

O'Sullivan, J. N. (2012). The burden of durable asset acquisition in growing populations. *Economic Affairs, 32*(1), 31–37.

O'Sullivan, J. N. (2013). *The cost of population growth in the UK.* UK: Population Matters. Retrieved from http://populationmatters.org/documents/cost_population_growth.pdf

O'Sullivan, J. (2016). Population projections: Recipes for action, or inaction? Population and Sustainability, *1*(1), 45–57. Retrieved from https://www.populationmatters.org/wp-content/uploads/2016/06/Population_and_Sustainability_Vol_1_No_1.pdf

Oregon Foundation for Reproductive Health. (2012). *One key question: "Would you like to become pregnant in the next year?"* Retrieved from http://www.onekeyquestion.org/

PAI, Pathfinder International, Sierra Club. (2015). *Building resilient communities: The PHE way*. Retrieved from http://womenatthecenter.org/wp-content/uploads/2015/07/Building-Resilient-Communities-The-PHE-Way.pdf.

Population Institute. (2015). *Demographic vulnerability: Where population growth poses the greatest challenges*. Retrieved from https://www.populationinstitute.org/demovulnerability/

PRB. (2011–2016). World population data sheet. Washington, DC: Population Reference Bureau. Retrieved from http://www.prb.org/Publications/Datasheets/2015/2015-world-population-datasheet.aspx

Riahi, K., van Vuuren, D. P., & Kriegler, E. (2017). The shared socioeconomic pathways and their energy, land use and greenhouse gas emissions implications: An overview. *Global Environmental Change, 42*, 153–168.

Rockström, J., Gaffney, O., Rogelj, J., Meinshausen, M., Nakicenovic, N., & Schellnhuber, H. J. (2017). A roadmap for rapid decarbonization. *Science, 355*(6331), 1269–1271. Retrieved from http://science.sciencemag.org/content/355/6331/1269.full

Ryerson, W. N. (2010). Population, the multiplier of everything else. In R. Heinberg & D. Lerch (Eds.), *The post carbon reader: Managing the 21st century's sustainability crises*. Healdsburg, CA: Watershed Media. Retrieved from http://www.postcarbonreader.com

Samir, K. C., & Lutz, W. (2014). The human core of the shared socioeconomic pathways: Population scenarios by age, sex and level of education for all countries to 2100. *Global Environmental Change, 42*, 181–192. https://doi.org/10.1016/j.gloenvcha.2014.06.004.

Schellnhuber, H. J., Rahmstorf, S., & Winkelmann, R. (2016). Why the right climate target was agreed in Paris. *Nature Climate Change, 6*, 649–653.

Schleussner, C.-F., Rogelj, J., Schaeffer, M., Lissner, T., Licker, R., Fischer, E. M., et al. (2016). Science and policy characteristics of the Paris agreement temperature goal. *Nature Climate Change, 6*, 827. https://doi.org/10.1038/NCLIMATE3096.

Searchinger, T., Hanson, C., Waite, R., Lipinski, B., Leeson, G., & Harper, S. (2013). *Achieving replacement level fertility* (World Resources Institute working paper) Instalment 3 of "Creating a Sustainable Food Future". Retrieved from http://www.wri.org/publication/achieving-replacement-level-fertility

Sinding, S. W. (2009). Population poverty and economic development. *Philosophical Transactions of the Royal Society of London. Series B, Biological Sciences, 364*, 3023–3030. https://doi.org/10.1098/rstb.2009.0145.

Spratt, D. (2016). *Climate reality check: After Paris, Counting the cost*. Breakthrough, March 2016. Retrieved from http://www.breakthroughonline.org.au/#!papers/cxeo

Starbird, E., Norton, M., & Marcus, R. (2016). Investing in family planning: Key to achieving the sustainable development goals. *Global Health, Science and Practice, 4*, 191. https://doi.org/10.9745/GHSP-D-15-00374.

Tavernise, S. (2015, July 5). Colorado's effort against teenage pregnancies is a startling success. *The New York Times*. Retrieved from https://www.nytimes.com/2015/07/06/science/colorados-push-against-teenage-pregnancies-is-a-startling-success.html

UN Economic and Social Council. (2010). *Report of the Secretary-General on the flow of financial resources for assisting in the implementation of the Programme of Action of the International Conference on Population and Development* (E/CN.9/2010/5). Retrieved from http://www.un.org/en/development/desa/population/documents/cpd-report/index.shtml

UNDESA. (2004). *World population in 2300*. Population Division, United Nations Department of Economic and Social Affairs. Retrieved from http://www.un.org/esa/population/publications/longrange2/WorldPop2300final.pdf

UNDESA. (2011). *World population prospects, the 2010 revision*. Population Division, United Nations Department of Economic and Social Affairs. Retrieved from http://www.un.org/en/development/desa/publications/world-population-prospects-the-2010-revision.html

UNDESA. (2013). *World population prospects, the 2012 revision*. Population Division, United Nations Department of Economic and Social Affairs. Retrieved from http://www.un.org/en/development/desa/publications/world-population-prospects-the-2012-revision.html

UNDESA. (2015) *World population prospects, the 2015 revision*. Population Division, United Nations Department of Economic and Social Affairs. Retrieved from https://esa.un.org/unpd/wpp/Graphs/Probabilistic/POP/TOT/

UNFPA. (1994). *Programme of action: Adopted at the international conference on population and development*. Cairo, Egypt: United Nations Population Fund (UNFPA).

van Vuuren, D. P., Kriegler, E., O'Neill, B., Ebi, K., Riahi, K., Carter, T. R., et al. (2014). A new scenario framework for climate change research: Scenario matrix architecture. *Climatic Change Special Issue, 122*(3), 373–386. https://doi.org/10.1007/s10584-013-0906-1.

Vörösmarty, C. J., Green, P., Salisbury, J., & Lammers, R. B. (2000). Global water resources: Vulnerability from climate change and population growth. *Science, 289*, 284–288.

Wheeler, D., & Hammer, D. (2010). *The economics of population policy for carbon emissions reduction in developing countries* (Center for Global Development, Working Paper 229, November 2010). Retrieved from http://www.cgdev.org/publication/economics-population-policy-carbon-emissions-reduction-developing-countries-working

Wittgenstein Centre for Demography and Global Human Capital. (2015). *Wittgenstein Centre Data Explorer Version 1.2*. Retrieved from www.wittgensteincentre.org/dataexplorer

Worldwatch Institute. (2015) *Vital Signs report: Food trade and self-sufficiency*. Retrieved from http://vitalsigns.worldwatch.org/vs-trend/food-trade-and-self-sufficiency

Young, M. H. Mogelgaard, K., & Hardee, K. (2009). *Projecting population, projecting climate change: population in IPCC Scenarios* (PAI Working Paper WP09–02). Population Action International. Retrieved from https://www.researchgate.net/publication/237249604_Projecting_Population_Projecting_Climate_Change_Population_in_IPCC_Scenarios

Zeng, N., Neelin, J. D., Lau, K. M., & Tucker, C. J. (1999). Enhancement of interdecadal climate variability in the Sahel by vegetation interaction. *Science, 286*, 1537–1540.

Part III
Major Challenges Towards a Sustainable Future: Case Studies from Asia

Chapter 8
From "Harmony" to a "Dream": China's Evolving Position on Climate Change

Paul Howard

Abstract The 2015 UN Climate Change Conference in Paris (COP21) saw China take a prominent role in conjunction with many developed countries in pushing for greater cuts in emissions and giving impetus to a push towards renewable energy. This position was in contrast to COP15 in Copenhagen in 2009, where China had stressed the distinction between the needs of developing countries to have greater flexibility in the shift away from fossil fuels. Indeed, in the intervening years between COP15 and COP21, China shifted to become a driver of ambitious targets and emissions goals.

China's leading role at COP21 in Paris reflected a domestic push to transform their economy to a focus on "green" energy as an integral part of economic strategy. The governing Chinese Communist Party is focused on maturing the Chinese economy to develop a larger service sector. Consequently, the government is shifting the economy away from labour-intensive industries towards high-tech industries such as the renewable sector. Along with the pure economic rationale underpinning the shift to green energy, there is also the opportunity to reinforce the government's legitimacy as many younger Chinese place greater onus on the government to ensure improved social and environmental outcomes. In short, a large-scale shift to renewables suits the Chinese Communist Party's objectives, in both economic and political terms.

Keywords China • Climate • Emissions • Renewables • Energy • COP21 • Economy • Wind • Solar

8.1 Introduction

At the 2009 United Nations Climate Change Conference in Copenhagen (COP15), much of the debate was framed around the notion of the developing world being forced to commit to targets imposed by the developed world, specifically the USA

P. Howard (✉)
Department of International Business and Asian Studies, Griffith University,
Brisbane, QLD, Australia
e-mail: paul.howard@griffith.edu.au

and its allies. The 2015 Paris Conference (COP21), though, saw a far more positive atmosphere with major states, including China and the USA, ultimately becoming signatories to the Paris Agreement to embark on measures to further reduce emissions. In the years between COP15 and COP21, the political focus of the Chinese Government had shifted more decisively towards transforming the national economy to one focused on elaborate manufactures, technology and an expanding service sector.

This transformation saw the government deeply engaged in developing renewable energy as a core of this "new" economy. This focus on economic transformation had come amid the orderly handover of power from Hu Jintao to Xi Jinping. Hu Jintao's regime had been characterised by an explicit in-principle adherence to "scientific development". This was juxtapositioned with the broader pursuit of attaining a "harmonious society" while Xi Jinping has promoted the concept of "China Dream". While these concepts are essentially (and arguably deliberately) vague, this evolving political narrative frames official policy. China's leading role at COP21 in Paris was emblematic of this evolution and underpins the strengthened political emphasis on renewable energy and the move towards the green economy.

The move to renewable energy for the Chinese Communist Party (CCP) is driven by far more than a need to fulfil quantified targets set by COP21. It is a happy coincidence that the global push for "greener" forms of energies dovetails with the PRC (People's Republic of China) Government's economic aims and circumstances. China had seen consistently high rates of GDP growth (9–10%) from the beginning of the reform period. More recently, however, economic growth has moderated and the government has sought to manage the shift to a lower but more sustainable rate of growth. Along with this moderating economic growth, the PRC Government is overseeing a shift in economic activity towards that of a mature economy with a larger service sector. As part of that transition, high-tech industries are favoured over labour-intensive manufacturing. The renewable energy sector offers an opportunity for such hi-tech development and service sector employment.

Along with the pure economic rationale underpinning the shift to "green" energy, there is also the opportunity to reinforce the government's legitimacy, particularly in the eyes of younger Chinese. From the reform period onwards, legitimacy had been premised on continued economic growth and development. In more recent times, the terms of political legitimacy have started to change in terms of perceptions of many younger Chinese. This may be particularly apt for those Chinese who have experienced relative prosperity in their formative years (Inglehart, 1997). Environmental concerns are core to issues relating to quality of life over pure economic performance. Consequently, the shift to renewable energy fits neatly into the CCP's appeal to many younger Chinese and potentially underpins the party's legitimacy.

This chapter is exploratory in nature with the overarching purpose being to shed light on the evolving narrative surrounding China's position on climate change and the rationale underpinning it. In doing so, the aim is to highlight the economically pragmatic case for China's policy shift towards the adoption of renewable energy. To achieve this aim, this work has primarily used official government publications, speeches and data in tandem with data from other sources, including international organisations. It is hoped that this approach will allow for a balanced analysis.

8.2 From Copenhagen to Paris: COP15 and COP21

The importance of creating a "greener" economy for both economic and sociopolitical reasons has been underscored by the Chinese Government's approach to climate change, both implicitly and explicitly. For example, the 13th Five-Year Plan (which covers 2016–2020) stated that the PRC Government would "take active steps to control carbon emissions, fulfil our commitments for emissions reduction, increase our capability to adapt to climate change, and fully participate in global climate governance" (NDRC, 2016). This broad commitment was supplemented with concrete measures, such as monitoring polluting industries and penalising and blacklisting polluters that fail to meet or respond to stipulated targets, and a specific commitment to controlling "total emissions of volatile organic compounds in chief regions and industries to bring about a nationwide drop in total emissions of over 10%" (NDRC, 2016).

While there is evidence of China's domestic efforts to curb pollution and embrace renewable energy, there is a corresponding story of China's evolution in its international engagement on environmental issues. This political evolution is well illustrated by the clear shift in the Chinese position from negotiations at COP15 in Copenhagen to COP21 in Paris. In those 6 years, China moved from an environmentally defensive stance to one of effectively taking a leading role in supporting and even bettering ambitious targets.

8.2.1 Copenhagen (COP15)

China's position in Copenhagen in 2009 was emblematic of the developing versus developed world argument. Indeed, as the largest developing economy, its reluctance to follow the lead of the USA and other powers provided legitimacy to other developing countries in their similar stance against the imposition of ambitious emissions targets. Similarly, one could argue that China's role at the 2015 Paris climate talks was a likely signpost for other developing countries. "Like other countries, China has to date followed a pattern of 'grow first, clean up later'. Yet very quickly it has recognised the dangers and drawbacks of such a policy and has been pouring money into clean energy and other innovations it hopes will provide green growth. In that, it may prove a model for other fast-developing countries" (The Economist, 2015).

The official position of the Chinese Government was explicitly outlined in the document Implementation of the Bali Roadmap: China's Position on the Copenhagen Climate Change Conference (PMPRCUN, 2009), which specified four guiding principles. The second of these guiding principles was "the principle of common but differentiated responsibilities". This was extrapolated as meaning that:

> Developed countries shall take responsibility for their historical cumulative emissions and current high per capita emissions to change their unsustainable way of life and to substantially reduce their emissions and, at the same time, to provide financial support and transfer technology to developing countries. Developing countries will, in pursuing economic development and poverty eradication, take proactive measures to adapt to and mitigate climate change.

The theme of "differentiated responsibilities" was continued in the third and fourth principles, the former asserting (PMRCUN, 2009) that "Within the overall framework of sustainable development, economic development, poverty eradication and climate protection should be considered in a holistic and integrated manner so as to reach a win–win solution and to ensure developing countries to secure their right to development". The fourth guiding principle called for "The fulfilment of commitments by developed countries to provide financing, technology transfer and capacity building support to developing countries is a condition sine qua non for developing countries to effectively mitigate and adapt to climate change".

8.2.2 Paris Agreement (COP21)

In the lead-up to COP21, China had implemented major domestic policy shifts and had begun to embrace renewable energy on an unprecedented scale.

> China is leading the way. By placing the emphasis on production scale and market growth, it is contributing more than any other country to a climate-change solution. Its build-up of renewable-energy systems at serious scale is driving cost reductions that will make water, wind and solar power accessible to all. (Mathews & Tan, 2014, p. 168)

Following from this domestic shift in policy, COP21 saw China's rhetoric no longer revolving predominantly around the familiar theme of differentiated responsibilities. That tune had instead been replaced with an almost universal outlook to the importance of all nations playing their part to address climate change and ensure the long-term environmental health of the planet.

8.2.2.1 Commitments

The Paris Agreement incorporates both qualitative and quantitative commitments. For example, Article 2.1 of the Agreement (UNFCC, 2015) explicitly states that "in the context of sustainable development and efforts to eradicate poverty", signatories commit to:

(a) Holding the increase in the global average temperature to well below 2 °C above pre-industrial levels and pursuing efforts to limit the temperature increase to 1.5 °C above pre-industrial levels, recognizing that this would significantly reduce the risks and impacts of climate change.

Along with this specific target, the Agreement also contains some very broad-ranging goals and objectives. For example, parts b and c (respectively) of that article involve qualitative measures to increase the "ability to adapt to the adverse impacts of climate change and foster climate resilience and low greenhouse gas emissions development, in a manner that does not threaten food production" and to make "finance flows consistent with a pathway towards low greenhouse gas emissions and climate-resilient development" (UNFCC, 2015).

The Agreement also specifically highlights the variance in roles of developed and developing economies. For example, Article 9.1 states that "Developed country Parties shall provide financial resources to assist developing country Parties with respect to both mitigation and adaptation in continuation of their existing obligations under the Convention" (UNFCC, 2015). China had committed to meet and even exceed these targets, something that perhaps sealed its (perhaps reluctant) position as the "flagship" of the developing world. Moreover, China's growing economic and political influence in global affairs is more firmly casting the importance of its leadership more generally. Green and Stern (2015, pp. 43–44) outline several specific reasons underscoring the importance of China's actions:

1. China's sheer size—geographically, demographically, economically and in terms of its energy use and greenhouse gas emissions—means China will always be a critical participant in global action on climate change.
2. China is seen by many developing countries as a model in the structure of economic growth and development, and as a leader in world economic affairs.
3. China influences politics in rich countries. There is a lack of understanding in rich countries about the measures China has already taken, and its future plans with regard to emissions.
4. China's strategy for reducing emissions can set an example for all countries.

It is the last of these points that is perhaps most noteworthy. For despite its domestic efforts to curb the polluting industry and emissions, during earlier negotiations at the United Nations Framework Convention on Climate Change, China had developed a "reputation as a 'laggard' or 'hard-liner'" (Held, Nag, & Roger, 2012, p. 29). In more recent years, however, China's role became more one of a developing country leader. Its place as the world's second largest economy and its growing role in international agreements and negotiations with developed countries has now placed it in a broader leadership position. In the wake of the 2016 UNFCC in Paris, UN Secretary-General Ban Ki-Moon asserted that China had "significantly accelerated political momentum towards the agreements rapid entry into force" (UN, 2016). China's role internationally, though, is both dependent on and reflective of domestic considerations. While the CCP's legitimacy is increasingly dependent on a broader range of factors, economic outcomes are still paramount.

8.3 Policy Shift: The Economic Imperative

8.3.1 Domestic Economic Transformation

Hilton and Kerr (2016) have asserted that the change in stance by China from COP15 to COP21 was driven by "China's shift to a 'new normal' model of economic development", that being one driven by an increased focus on good and services and slower growth. This, they argue, "helped pave the way" for China to work cooperatively with the USA. However, they add that the pivot was also facilitated by "external factors" such as "impressive French diplomacy" and a "shift to a bottom-up, voluntary approach to commitments" (Hilton & Kerr, 2016, p. 48).

Green and Stern (2015, p. 8) have identified three main reasons for the unsustainability of the "old model of growth":

1. Over-investment and diminishing returns and, specific to energy, over-capacity in energy intensive sectors.
2. Rising wages coupled with a falling proportion of workforce aged [defined by them as 16–60] population—need to shift to "higher value-added" industry.
3. Environmental degradation and "natural resource constraints" coupled with a dependence on energy imports which has led to rising economic costs.

Along with the unsustainability of the "old" growth mode, Green and Stern (2015, p. 8) have noted that "this phase of China's development also caused high growth in China's greenhouse gas emissions".

The picture of China's dependence on fossil fuels is an uneven one. A 2016 report found the highest levels of CO_2 emissions to be in the provinces of Inner Mongolia, Ningxia and Shanxi (Liu, 2016). The common thread with these provinces is the dependence on fossil fuels, particularly coal, by state-owned enterprises. This represents the old growth model, one which was driven by economic benefit with little or no consideration of environmental factors. In fairness, it should be added that these high per capita emission provinces export electricity and industrial goods to other provinces. Indeed, about 50% of Inner Mongolia's GDP is constituted by these inter-provincial exports of electricity and industrial goods (Liu, 2016, p. 10).

8.3.2 Investment

In many respects, this regional/provincial disparity of per capita emissions is perhaps more symptomatic of deeper causal factors, not least of which is the inconsistency of local governmental policy implementation and investment oversight. Zhang et al. (2017, p. 163) have asserted that:

> The current overcapacity in China's coal production has an obvious local government feature. As in the context of fiscal decentralization, the motivations for the central and local government to intervene in coal production investment are inconsistent. Local government and coal enterprise pursue maximum benefits, whereas the central government pursues the orderly development of the industry. Apparently, there are asymmetric information and benefits conflict of coal capacity investment among local government, coal enterprise and the central government.

This "inconsistency" and even disconnect between environmental regulation at the national level and local governmental actions in regard to coal investment and capacity is unsurprising. Of course, even in highly efficient developed economies, federal objectives run into local roadblocks and difficulties. In the case of environmental concerns, this may be due to electoral contestation and/or nongovernment organisation activism. In the case of the PRC, even in terms of formal policy and actions, there is the common instance of the broader objectives of national-level policy being inconsistent with local level needs or aims. Consequently, the issue of local implementation and actions being incongruent or even directly at odds with national policy may be extended to investment and capacity in the renewable energy sector as well. In other words, it would be short-sighted to assume this national–local disconnect was limited to coal capacity and investment.

Investment in the renewable energy sector has been given impetus by significant government-driven incentives. For example, qualifying technology enterprises are eligible for a reduced corporate tax rate of 15% (KPMG, 2015, p. 22). Solar energy and wind power are included in the eligible fields along with geothermal and biomaterial energy projects. There is also a 50% refund of VAT on wind power sales (KPMG, 2015). Along with this, there are tax benefits for sustainable technologies incorporating the renewable energy sector, some of which involve relatively complex tax benefits related to the clean development mechanism fund. Less specific subsidies are available for the "demonstration of important technologies for renewable and new energy", "development and utilization of renewable and new energy", "development of platforms for renewable and new energy" and other categories relevant to renewable energy "as approved by the State Council" (KPMG, 2015, p. 24).

At face value, the progress and goals in environmental management have been significant. It is pertinent to examine the actual economic commitment. In 2013, China produced more than double the renewable energy output of the USA (378 GW), with hydroelectric the main contributor and wind power making up the major share of the balance (Mathews & Tan, 2014). The desire of the central government to move to renewable and relatively "green" energy sources is matched by vested interests with strong political linkages. There is, moreover, a direct reluctance by many provincial and local level CCP officials to move away from coal, given its importance as a source of employment, particularly in Shanxi and Inner Mongolia, for example (The Economist, 2015).

According to HSBC, the PRC Government's expenditure on treating environmental pollution rose 3% year on year from 2014 to 2015, as compared to expenditure as a whole, which rose 1% in the same period (He, 2016). See Fig. 8.1.

As a percentage of GDP, total investment in the treatment of pollution rose slightly for the five-year period from 2009 to 2013. In 2009, the total investment was 1.54% of GDP, with the total investment in 2010 rising to 1.89% of GDP. By 2013, however, total investment was 1.67% of GDP (NBS, 2014). See Fig. 8.2.

8.3.3 Capacity and Growth

Given the explicit incentives for investment in renewable energy, it raises the question of the impact that these incentives have on actual outcomes. In that regard, China's changing power landscape is evidenced by the official data. According to

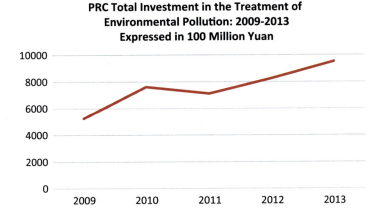

Fig. 8.1 PRC total investment in the treatment of environmental pollution (2009–2013) expressed in 100 Million yuan (Source: NBS, 2014)

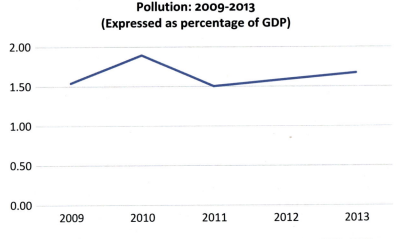

Fig. 8.2 PRC total investment in the treatment of environmental pollution (2009–2013) expressed as a percentage of GDP (Source: NBS, 2014)

China's National Bureau of Statistics, total power generation capacity in the PRC reached 1.5 million kW in 2015, a 10.5% increase on the 2014 figure (NBS, 2016). Disaggregating the data, in that 12-month period, the power generation capacity of thermal power rose by 7.8% to 990 million kW. As the largest renewable energy source, hydropower capacity rose by 4.9% in the period to 319 million kW and nuclear power capacity rose by a substantial 29.9% to 26 million kW (NBS, 2016). However, the greatest growth in power generation capacity was in the solar and wind power.

Of these two renewable sources, "grid-connected" wind power generating capacity increased by 33.5% to 129.34 million kW. However, by far the largest rise in generating capacity in the period was solar power (grid-connected) which saw a 73.7% increase in capacity, rising to 43.18 million kW by 2015 (NBS, 2016). In broad terms, China (2014) was the leading investor in renewable power and fuels followed by the USA, Japan, the United Kingdom and Germany. In terms of specific sectors, China had the largest wind power capacity (end of 2014) followed by the USA, Germany, Spain and India. In terms of solar energy capacity, China also led, followed by Japan, the USA, the United Kingdom and Germany (KPMG, 2015, p. 77).

In their analysis of the economic impact of "green energy" in the PRC, Dai, Xie, Xie, Liu, and Masu (2016) used a computable general equilibrium model to analyse the economic impacts under two scenarios: a reference scenario "assuming conventional development" of renewable energy (RE) and an "REmax scenario assuming large-scale RE development by tapping China's RE potential". The study concluded that under the REmax scenario (i.e. major investment in renewable energy) there would be relatively minimal negative macroeconomic impacts. At the same time, they asserted that there would be numerous economic benefits to the general economy with significant stimulatory impact on a variety of "upstream" industries, including machinery and electronics manufacturing and R&D (Dai et al., 2016).

Moreover, by 2050, Dai et al. (2016) estimate that non-fossil power industries could account for 3.4% of GDP and that 4.12 million jobs could be created. While they emphasise the net economic benefits of a large-scale shift to renewable energy, they also note that conventional energy industries would suffer. Consequently, Dai et al. (2016) note the need for the government to undertake "countermeasures" to manage the economic and labour shift from conventional energy industries to the renewable energy sector.

8.4 From Policy to Priority

Chen and Lees (2016, p. 582) have linked the PRC's political and economic structure to the "expansion of renewable energy in China" which they assert:

> … was characterised by the enhancement of central steering capacity, consistent with the developmental state paradigm. In the Chinese model, neither market efficiency nor increasing societal participation was a priority for restructuring the energy market. Instead, central government securitised the sector through re-centralisation and re-intervention, giving priority to political stability through its lead agency, the NDRC.

Shen (2016) has argued that the governance of China's energy sector has evolved over the last decade or so to fragment government institutions, something that mitigates the incidence of "institutional conflict". This fragmentation of institutions in the renewable energy sector has not led to "policy contradictions" often inherent in the Chinese context (Shen, 2016, p. 6), due to the variance in "decision-making power" of the respective government bodies. Among these bodies, the National

Energy Administration, as part of the National Development and Reform Commission (NDRC), has the institutional power of the NDRC and its direct linkages to the Chinese State Council behind it (Shen, 2016).

8.4.1 Five-Year Plans

This tight linkage between political process and environmental oversight is inherent in the government's official five-year plans. The 12th Five-Year Plan outlined a commitment to addressing climate change and developing the "green" economy. For example, Part IV of the plan concerned "green development, construct energy conservation and environment friendly society" (CBI China, 2011) which was fairly broad in its overarching aims.

> We will establish green and low carbon development ideas and focus on energy conservation and emission reduction, improve incentives and constraint mechanisms, and stimulate the establishment of resource-saving and environmentally friendly production and consumption to strengthen sustainable development and improve ecological standards.

While the document emphasised the need for international "cooperation" following the tenor of the Copenhagen negotiations 2 years earlier, the government continued to invoke the developed versus developing world dichotomy. "Adhering to the principle of common but differentiated responsibilities, we will actively participate in international negotiations and promote the establishment of a fair and reasonable international system for confronting climate change" (CBI China, 2011). Specific sections focused on controlling greenhouse emissions, increasing "adaptability to climate change", and "wide-ranging international cooperation" (CBI China, 2011). For the most part, though, goals concerned more aspirational aims than specific targets.

The 13th Five-Year Plan came in the wake of the Paris Agreement and, in accordance with that international Agreement, set ambitious domestic emissions targets. Specifically, the 13th Five-Year Plan set a target of a 48% drop in emissions from 2005 levels by 2020. This compared to the Paris commitment of a 40–45% reduction in emissions from 2005 levels by 2020 (Henderson, Song, & Joffe, 2016). Significantly, this is coupled with growth of greater than 6.5% per annum driven by economic transition as the service sector is to rise from 50.5% to 56% of China's GDP (Henderson et al., 2016).

8.4.2 Annual Reports (2009 and 2016)

While the five-year plans lay out the government's plan for the ensuing 5-year period, the annual reports on the "work of the government" provide the Premier an opportunity to espouse the government's achievements over the prior year and to outline the objectives for the following year. The 2009 and 2016 reports bookmark the period from prior to COP15 and after COP21, respectively. In the 2009 Report on the Work of the Government, references to emissions and other environmental matters essentially came under subsections 1 and 4 entitled "Improving and strengthening macro control and maintaining steady and rapid economic development" and "Accelerating transformation of the pattern of development and energetically promoting strategic economic restructuring", respectively (Wen, 2010).

The significance of this is that the explicit priority is framed as economic development with environmental considerations as a necessary consideration rather than being central to policy objectives. There was, nevertheless, specific attention given to environmental concerns, with the report outlining that 42.3 billion yuan had been allocated to "support development of ten key energy conservation projects and environmental protection facilities" (Wen, 2010). In the same period, "energy consumption per unit of GDP fell by 4.59% from the previous year; chemical oxygen demand fell by 4.42%; and sulfur dioxide emissions fell by 5.95%" (Wen, 2010). Premier Wen Jiabao (Wen, 2010) did give specific mention to climate change and sequestration:

> We built on the achievements in returning farmland to forests and restoring livestock pastures to grasslands, and carried out ecological conservation projects such as protecting virgin forests and developing the Sanjiangyuan Nature Reserve in Qinghai Province. We implemented the plan to prevent and control water pollution in major river valleys and regions and issued the white paper China's Policies and Actions for Addressing Climate Change.

In the March 2016 "Report on the Work of the Government", Premier Li Keqiang (2016) explicitly outlined the government's commitment to "step up environmental governance efforts and work to see breakthroughs in green development". This commitment was given, however, as taking "a path that leads to both economic development and environmental improvement" (Li Keqiang, 2016). In terms of specific quantifiable commitments for 2016, Premier Li (Keqiang, 2016) outlined that the government would "ensure" that:

- Chemical oxygen demand and ammonia nitrogen emissions are both reduced by 2%
- Sulphur dioxide and nitrogen oxide emissions are cut by 3%
- The density of fine particulate matter in key areas continues to fall

These specific goals were coupled with a government commitment to:

- Promote the reduction of emissions from the burning of coal and motor vehicles
- Work to promote cleaner and more efficient use of coal, ensure there is less use of untreated coal, do more to see coal substituted with electricity and natural gas, and upgrade coal-burning power plants nationwide to achieve ultra-low emissions and energy efficiency
- Move faster to shut down coal-fired boilers that do not meet compulsory standards
- Increase natural gas supply

- improve policy support for the development of wind, solar and biomass energy, and increase the proportion of clean energy in total energy consumption
- Encourage a whole range of forms of straw utilisation and restrict open burning
- Push for the use of automobile gasoline and diesel fuel that meet National-V standards nationwide and see that 3.8 million old or high-emission vehicles are taken off the roads
- Coordinate efforts to prevent and control air pollution in regions where it presents a problem
- Press ahead with the nationwide development and upgrading of urban sewage treatment facilities, and strengthen comprehensive efforts aimed at controlling agricultural pollution from non-point sources and improving the water environment in river basin areas
- Step up efforts to deal with industrial pollution at the source and conduct online monitoring of all polluting enterprises
- Strengthen environmental inspection and reward and punish as appropriate
- Ensure that the newly revised Environmental Protection Law is strictly enforced

Both the stating of these broader goals along with the more specific evaluation of performance in the annual reports is an opportunity for the government to highlight both achievements and commitments. Both the five-year plans and the annual reports are reported on in detail by state media and disseminated via official news channels, making them an important means of reinforcing the public perception of the CCP's performance.

8.5 Public Perception and Civil Society

Throughout the reform period and until relatively recently, the CCP's legitimacy had been based on pure economic performance. Decades of 9-10% per annum GDP growth enabled the CCP to continue the theme of economic development as core to policy and public outcomes in terms of improved standards of living, urban infrastructure and increasing personal wealth. In more recent times, however, there has been growing dissatisfaction among the Chinese about environmental problems, many of which were seen to be caused by rapid economic development. With a rising middle class, this shift may be linked to Inglehart's (1997) notion of "materialist/postmaterialist value change", a concept underpinned by both a "scarcity" hypothesis and a "socialisation" hypothesis. The former essentially asserts that an individual "places the greatest subjective value on those things that are in short supply". The latter posits that there is a time lag between an individual adjusting to socioeconomic transformation as "to a large extent, one's basic values reflect the conditions that prevailed during one's pre-adult years" (Inglehart, 1997, p. 132).

The reform period, which began in the late 1970s, may be seen as ushering in major socioeconomic transformation, which could ultimately inform greater participation in civil society (Howard, 2011). The "time lag" Inglehart (1997) described could be transposed to the Chinese case such that many who benefited from economic development in their formative years during, say, the 1990s onwards have been subject to that "postmaterialist value change". That value change may be reflected in the results of a 2016 Ipsos poll of 3000 urban Chinese that found that 90% of respondents were concerned with environmental pollution, with 96% agreeing with the idea that renewables could help address climate change (Murray, 2016). In addition, 92.6% of respondents accepted the idea of paying more for electricity from renewable sources (Murray, 2016).

The changing attitudes towards environmental integrity have corresponded to the development of large numbers of environmental groups and nongovernment organisations, particularly at the "grassroots" level. Grassroots civil society may be constrained by government regulations and still lack mass participation. However, there is a sense that mainstream support for activism on environmental matters is growing, particularly among the young. In their study of protests to save London Plane trees in Nanjing, Grano and Zhang (Grano & Zhang, 2016, p. 177) have noted the importance of even small victories by environmental activists for broader shifts in public sentiment:

> And even if such participatory channels are not yet fully functioning or developed in the People's Republic of China, the fact that dozens of trees were saved from destruction should not be taken lightly in a country where thousands of hectares of forests and farmland are deforested and cleared every day under the all-embracing imperative of "economic development first". Such progress indicates the arrival and gradual establishment of what Ronald Inglehart named "post-materialist values" in China's urban centres.

Such wins for activists are complemented by increasing recognition by senior figures within the CCP of the need for immediate implementation of measures to solve China's problems with water and air pollution. Indeed, Grano and Zhang (Grano & Zhang, 2016, p. 166) note that from around 2013 onwards, many "high-ranking party cadres" have given speeches on environmental issues. This entrance of environmental issues into the official political narrative challenges the commonly drawn contrast between liberal democracies and authoritarian states, with the former being where the free flow of ideas is encouraged versus the latter where the state manages both policy and implementation without objective interference. The inference is that the latter will follow rather than lead in developing and implementing policy that confront issues such as climate change.

8.6 Challenges

8.6.1 *Economic Development*

The political impetus to shift to renewable energy has come amid a slowing of growth in the Chinese economy. The capacity needs of a burgeoning economy provide impetus to the uptake of renewable energy and, conversely, an economic slowdown may potentially impact the renewable sector. China's economic (GDP) growth had been sustained at around 9–10% per annum from the reform period in the late 1970s until recently. In more recent years, China's economic growth began to moderate, with growth slowing to just under 7% per annum in 2015 and 2016 (World Bank, 2017). While numerous factors may be at play in determining electricity demand growth, there has been a correlation between electricity demands in China and the slowdown in economic growth.

For example, after decades of significant growth in electricity demand, 2015 saw that growth drop to just 0.5–1% (Slater, 2017). The 5% growth recorded in 2016 was primarily due to increased industrial output due to government stimulus of the property sector, and Slater (2017) observes that "as the government aims to cool the property market over coming months, this effect is likely to disappear". This does pose significant issues for the government in its support of the renewable energy sector. Shen (2016) has asserted that there has been substantial curtailment of wind farm activity since 2014 due to energy grid oversupply. Shen cites the testimony of one interviewee (Interviewee no. 8, December 2014 cited in Shen 2016, pp. 18–19):

> Before you invest, they [the local governments] gave you all sorts of promises: land, tax free, clearances and permits, etc. But when your investment got underway, you were then told that they would not help you sell the electricity—worse than that, you have to give way to fossil fuel power, which is essentially violating the Renewable Energy Law of the People's Republic of China.

8.6.2 Renewable Energy law

The intent of the Renewable Energy Law is explicitly stated as follows in Article 1 (MOFCOM, 2013):

> This Law is enacted for the purpose of promoting the development and utilization of renewable energy, increasing the supply of energy, improving the structure of energy, safeguarding the safety of energy, protecting [the]environment and realizing a sustainable economic and social development.

The assertion that investors are finding they are held up by "red tape" and yielding to fossil fuel energy does at face value seem contrary to the broad intent of the law but, more specifically, is arguably at odds with the statement of priority given in Article 4 of the Renewable Energy Law which states that:

> The state shall give priority to the development and utilization of renewable energy in energy development and promote the establishment and development of the renewable energy market by setting an overall target for the development and utilization of renewable energy and adopting corresponding measures.

> The state shall encourage economic subjects of different ownership to participate in the development and utilization of renewable energy and shall protect the legitimate rights and interests of those who develop and utilize renewable energy.

This seeming contradiction between stated intent under the Renewable Energy Law and real outcomes for some suppliers may point more to changes in the economic context. For example, the Renewable Energy Law was promulgated in 2013 when both broader economic conditions and the development of the renewable energy grid were very different from their respective subsequent states. China's GDP growth had tapered from around 9% in 2011 to 7.15% in 2013 (World Bank, 2016). In that same period, grid-connected renewable energy increased significantly. The slowing economy and a commensurate slowdown in the growth of energy requirements served to exacerbate the oversupply issue.

8.6.3 Wind Power

One of the challenges for the central government is that of providing impetus to renewable energy producers while still moving towards a true market-oriented pricing structure. For example, since 1986, the wind power industry has been through several pricing phases. The first was a "demonstration phase", from 1986 till 1993, in which the tariff structure for wind power was the same as that for coal power (around 0.3 yuan per KWh) (An, Lin, Zhou, & Zhou, 2015).

During the following "industrialisation phase", from 1993 to 2004, pricing was determined by mutual agreement via negotiation between power grid operators and the respective wind power companies. Pricing varied widely from 0.38 yuan per KWh up to as much as 1.2 yuan per KWh. From 2003 to 2009, the wind power industry entered a "scale-up and localisation phase" during which the NDRC oversaw the inception of "concession bidding". This phase saw the wind power industry enter a more market-driven pricing paradigm. Since 2009, according to An et al. (2015), there has been a combination of "approved tariff" pricing and pricing by tender.

8.6.4 Coal Price Trap

The move to a market-driven pricing structure for wind power is also coupled to market forces with respect to the price of fossil fuels, as is the case with the "coal price trap" scenario. Qi et al. (2016, p. 1) have argued that China hit peak coal consumption in 2013–2014. Coal consumption dropped by 2.9% in 2014 which was followed by a further 3.6% drop in 2015. This indicates not only that China hit peak consumption in 2013–2014 but that the negative growth in coal usage was accelerating. This in and of itself may be a positive development, but the challenge is then to avoid the trap of lower demand leading to lower prices which in turn leads to a reuptake of coal intensive energy supply (Qi et al., 2016). This potential lower price leading to a resurgence on fossil fuel usage underscores the need for government policy to ensure that this scenario does not transpire.

8.6.5 International Political Developments

In tandem with, and to some degree connected to, domestic concerns, there is the potential for any change in the international political economy to disrupt China's internal plans. Along with economic risks, there is also the dependence of international agreements on both economic events and political changes. For example, China and the USA both ratified the Paris Agreement on 3 September 2016 (UNFCC, 2016). However, the US election of November 2016 saw the prospect of a new incoming administration, which cast some uncertainty over the USA's role. However, despite this uncertainty, China remained publicly committed to its low emission targets. Kahn (2016) has suggested three main reasons for China's leadership's resoluteness to maintain an aggressive emissions reduction strategy:

1. To improve the quality of life in their nation's cities by reducing air pollution
2. To win large shares of promising export markets for green technologies
3. To increase China's "soft power" in international relations

With this aim of increasing its "soft power" internationally, China's role in addressing climate change becomes even more pertinent. As the world's second largest national economy, any reversion of US commitments to climate change agreements may be seen as an opportunity for Beijing to increase China's standing among developed countries by filling any "vacuum" that may be created. Given China's push for renewable energy domestically and its already growing role in international negotiations, China would be well positioned to capitalise and potentially garner greater support from many developed countries.

8.7 Conclusion

There are certainly numerous potential challenges to China's ongoing commitment to renewable energy, both domestically and internationally. For example, slowing but relatively stable economic growth may not be a direct threat to the Chinese Government's goals with respect to renewable energy. A significant slowdown either domestically or internationally could weaken the case for a rapid shift to alternative energy sources. Similarly, major changes to the international political economy or willingness by other nation-states to participate in the rollout of renewable sources could also impact China's ability to meet agreed targets.

Importantly though, the CCP's legitimacy had until recent years been premised, almost exclusively, on economic performance. With generational change has come a commensurate change in the basis of the party's legitimacy, a legitimacy that is now increasingly likely to be underpinned by "quality of life" and, with that, environmental governance. The growing focus on environmental issues has seen increasing participation in civil society, particularly among younger Chinese. It is in this context of generational change and shift in focus that the Chinese Government has enacted policy to encourage the development of the "green economy" with specific policy measures to promote growth in the renewable energy sector.

While the changing terms of legitimacy may be coincident with the move to a "green" economy, there is also an overarching economic imperative. The PRC Government has stated its desire for the country to move towards a larger service sector, something that could be encouraged by the development of the renewable energy sector.

Both the changing terms of political legitimacy and the economic imperative to develop the high-tech industry and the service sector generally provide clear motivation for the government to embrace an aggressive climate change strategy. These dual motivations have been manifested in China's change in international actions as evidenced by the very different outcomes of COP15 and COP21. The Chinese Government now has compelling and logical reasons to continue to take the initiative in setting ambitious environmental targets, both domestically and with its international partners.

This raises the question of how such momentum towards renewable energy sources will impact China's trading partners. Countries such as Australia, that export commodities to China, will need to be responsive to potential changes in market conditions. Indeed, China's policy in the area of renewable energy may have substantial implications for the approach taken by other countries in the region. Further, China's growing geopolitical influence may extend such implications to countries in other regions, particularly developing countries that are in any way tied to China politically or economically.

This chapter is, as was outlined earlier, essentially exploratory in nature. It was not intended as a definitive assessment of China's climate change or renewable energy policies. There exists, though, great scope for substantive quantitative and qualitative research to better understand China's evolving position with regard to both climate change and the unprecedented scale of the shift towards renewable energy. Indeed, those trading partners that best understand the motivations and economic underpinnings of China's shift to renewable energy will be best placed to pre-empt and respond to resultant changes in market demands.

References

An, B., Lin, W., Zhou, A., & Zhou W. (2015).*China's market-oriented reforms in the energy and environmental sectors* (Pacific Energy Summit 2015 Working Paper).

CBI China. (2011). *12th five-year plan* (English translation). Available from http://www.cbi-china.org.cn/cbichina/upload/fckeditor/Full%20Translation%20of%20the%2012th%20Five-Year%20Plan.pdf

Chen, G., & Lees, C. (2016). Growing China's renewables sector: A developmental state approach. *New Political Economy, 21*(6), 574–586.

Dai, H., Xie, X., Xie, Y., Liu, J., & Masu, T. (2016). Green growth: The economic impacts of large-scale renewable energy development in China. *Applied Energy, 162*, 435–449.

Grano, S., & Zhang, Y. (2016). New channels for popular participation in China: The case of an environmental protection movement in Nanjing. *China Information, 30*(2), 165–187.

Green, F., & Stern, N. (2015). *China's "new normal": structural change, better growth, and peak emissions*. Centre for Climate Change Economics and Policy (CCCEP) and Grantham Research Institute on Climate Change and the Environment, July 2015 Policy Brief.

He, L. (2016, February 4). China's environmental service sector to benefit from big government spending. *South China Morning Post (SCMP)*. Retrieved July 14, 2016 from http://www.scmp.com/business/markets/article/1909228/chinas-environmental-service-sector-benefit-big-government-spending

Held, D., Nag, E., & Roger, C. (2012). *The governance of climate change in developing countries: A Report on International and Domestic Climate Change Politics in China, Brazil, Ethiopia and Tuvalu. LSE-AFD Climate Governance Programme* (Savoir No 15, October 2012). Retrieved April 12, 2017 from http://www.afd.fr/jahia/webdav/site/afd/shared/PUBLICATIONS/RECHERCHE/Scientifiques/A-savoir/15-VA-A-Savoir.pdf

Henderson, G., Song, R., & Joffe, P. (2016). *5 Questions: What Does China's New Five-Year Plan Mean for Climate Action?* World Resources Institute. Retrieved September 28, 2016 from http://www.wri.org/blog/2016/03/5-questions-what-does-chinas-new-five-year-plan-mean-climate-action

Hilton, I., & Kerr, O. (2016). The Paris agreement: China's 'new normal' role in international climate negotiations. *Climate Policy, 17*(1), 48–58.

Howard, P. (2011). 'harmony' in China's climate change policy. In M. Hossain & E. Selvanthan (Eds.), *Climate change and growth in Asia* (pp. 177–192). Cheltenham: Edward Elgar.

Inglehart, R. (1997). *Modernization and postmodernization*. Princeton, NJ: Princeton University Press.

Kahn, M. (2016). For China, climate change is no hoax—it's a business and political opportunity. *The Conversation*. Retrieved December 11, 2016 from http://theconversation.com/for-china-climate-change-is-no-hoax-its-a-business-and-political-opportunity-69191

Keqiang, L. (2016). Report on the work of the government. Retrieved April 4, 2016 from http://news.Xinhuanet.Com/english/china/2016-03/17/c_135198880_2.Htm.

KPMG. (2015). *Taxes and incentives for renewable energy*. KPMG International. Retrieved from https://assets.kpmg.com/content/dam/kpmg/pdf/2015/09/taxes-and-incentives-2015-web-v2.pdf

Liu, Z. (2016). *China's carbon emissions report 2016*. Cambridge, MA: Harvard Belfer Center for Science and International Affairs.

Mathews, J., & Tan, H. (2014). Manufacture renewables to build energy security. *Nature, 513*, 166–168.

MOFCOM (Ministry of Commerce People's Republic of China). (2013). *Renewable Energy Law of the People's Republic of China*. Retrieved August 12, 2016 from http://englisfh.mofcom.gov.cn/article/policyrelease/Businessregulations/201312/20131200432160

Murray, J.(2016, August 31). Chinese public overwhelmingly backs renewables push. *Business Green*. Retrieved November 12, 2016 from http://www.businessgreen.com/bg/news/2469252/chinese-public-overwhelmingly-backs-renewables-push.

NBS. (2016). *Statistical Communiqué of the People's Republic of China on the 2015 National Economic and Social Development*. Retrieved February 30, 2016 from http://www.stats.gov.cn/english/PressRelease/201602/t20160229_1324019.html

NBS (National Bureau Statistics of China). (2014). *China statistical yearbook 2014*. Retrieved from http://www.stats.gov.cn/tjsj/ndsj/2014/indexeh.htm.

NDRC (National Development and Reform Commission). (2016). The 13th Five Year Plan for economic and social development of the People's Republic of China (2016–2020), Retrieved April 12, 2017 from http://en.ndrc.gov.cn/newsrelease/201612/P020161207645765233498.pdf

PMPRCUN (Permanent Mission of the Peoples' Republic of China to the UN). (2009). *Implementation of the Bali Roadmap: China's Position on the Copenhagen Climate Change Conference May 20, 2009*. Retrieved May 20, 2016 from http://www.china-un.org/eng/chinaandun/economicdevelopment/climatechange/t568959.htm

Qi, Y., Stern, N., Wu, T., Lu, J., & Green, F. (2016). China's post-coal growth. *Nature Geoscience, 9*, 564–566.

Shen, W. (2016). A new era for China's renewable energy development? External shocks, internal struggles and policy changes. (Evidence Report No 196). *Rising powers in international development*. Institute of Development Studies.

Slater, H. (2017).China's new lacklustre renewable energy target. *East Asia Forum*. Retrieved March 2, 2017 from http://www.eastasiaforum.org/2017/03/01/chinas-new-lacklustre-renewable-energy-targets/

The Economist (from the print edition). (2015, December 5). *Raise the Green Lanterns*. Retrieved May 14, 2016 from http://www.economist.com/news/china/21679500-china-using-climate-policy-push-through-domestic-reforms-raise-green-lanterns

UN (United Nations). (2016). In China, Ban highlights country's leadership on sustainable development, climate change. *UN Sustainable Development Blog*. Retrieved from http://www.un.org/sustainabledevelopment/blog/2016/07/in-china-ban-highlights-countrys-leadership-on-sustainable-development-climate-change/

UNFCC (United Framework Convention on Climate Change). (2015). *Paris Agreement, Conference of the Parties Twenty-first session Paris, 30 November to 11 December 2015*. Retrieved from https://unfccc.int/resource/docs/2015/cop21/eng/l09r01.pdf.

UNFCC (United Framework Convention on Climate Change). (2016). *Paris Agreement: Statement of ratification*. Retrieved from http://unfccc.int/paris_agreement/items/9444.php.

Wen, J. (2010). *Report on the work of the Government 2009*. Retrieved June 30, 2012 from http://www.npc.gov.cn/englishnpc/Special_11_5/2010-03/03/content_1690628.htm

World Bank. (2016). *World development indicators: China GDP per capita*. Retrieved from http://databank.worldbank.org/data/reports.aspx?source=2&series=NY.GDP.PCAP.KD.ZG&country=CHN#

World Bank. (2017). *GDP growth: China*. Retrieved from http://data.worldbank.org/indicator/NY.GDP.MKTP.KD.ZG?end=2015&locations=CN&start=2015&view=map

Zhang, Y., Zhang, M., Liu, Y., & Nie, R. (2017). Enterprise investment, local government intervention and coal. *Energy Policy, 101*, 162–169.

Chapter 9
COP21 and India's Intended Nationally Determined Contribution Mitigation Strategy

Ranajit Chakrabarty and Somarata Chakraborty

Abstract One of the greatest anxieties facing humanity today is anthropogenic climate change and its global impacts due to global warming. Since 1992, the United Nations Framework Convention on Climate Change Conference of the Parties (COP) has been seeking global solutions to this issue. In 2015, at COP21, there was a universal agreement to reduce global emissions to limit global warming to below 2 °C. In this so-called Paris Agreement, nations agreed to rapidly reduce their carbon emissions through a process of submitting intended nationally determined contributions (INDCs) to mitigate climate change. In Sect. 9.1, global warming and the development of COP leading to COP21 are discussed. Section 9.2 presents India's position on the environment and its overall growth, and, on that basis, India's INDC. In Sect. 9.3, mitigation efforts for meeting the INDC and the stress on renewable energy are given, as it plays a significant role in the mitigation strategy. Section 9.4 discusses the different barriers that the mitigation effort is facing and the problem of demonetisation of the high-value notes in India. In Sect. 9.5, we conclude with the impediments that arise due to issues such as demonetisation of Indian currency, oil price shock and US policy.

Keywords UNFCCC • COP21 • INDC • GCF • Demonetisation

R. Chakrabarty (✉)
Department of Business Management, Calcutta University, Kolkata, India
e-mail: rchakrabarty4@hotmail.com

S. Chakraborty
Calcutta University, Kolkata, India

9.1 Introduction

9.1.1 Global Warming

It is now an established fact that climate change has affected every continent, from the equator to the poles, from the mountains to the coasts. In small island developing states, arable land, water resources and biodiversity are already under pressure from sea level rise. Increases in population and the unsustainable use of available natural resources add further problems. Tropical storms and cyclones cause storm surges, coral bleaching, inundation of land and coastal and soil erosion with resulting high-cost damage to socioeconomic and cultural infrastructure (see IPCC, 2014).

Due to industrialisation, the burning of fossil fuels, such as oil and coal (which emit heat-absorbing gases, such as carbon dioxide (CO_2), methane, nitrous oxide and chlorofluorocarbons), continues to increase the atmospheric temperature. Earth's climate is expected to warm even more rapidly during this century. The concentration of CO_2 in the atmosphere is increasing at a rate of about 1–2 parts per million (ppm) per year and it is currently around 390 ppm from the pre-industrial level of 280 ppm (see IPCC, 2014).

Developing countries are more vulnerable to climate risks because of limited capacity to adapt to rising sea levels or recover from associated losses (UNFCCC, 2007). The Bali Action Plan (see UNFCCC, 2008) and the ongoing climate change negotiations address enhanced action on adaptation, including the urgent and immediate needs of developing countries that are particularly vulnerable (UNFCCC, 2008). The least-developed countries and small island developing states are highly vulnerable to the impacts of climate change and are already feeling its impacts (UNFCCC, 2007) However, their contribution to global warming is markedly different from that of developed countries (Table 9.1).

9.1.2 The United Nations Framework Convention on Climate Change (UNFCCC

According to the Global Footprint Network (2016) today humanity uses the equivalent of 1.7 planets to provide the resources we use and absorb our waste. This means it now takes the Earth one year and 6 months to regenerate what we use in a year. Moderate UN scenarios suggest that if current population and consumption trends continue, by the 2030s, we will need the equivalent of two Earths to support us. And of course, we only have one.

As a consequence of the recognition of this situation, in 1988 the Inter-Governmental Panel on Climate Change (IPCC) was established by the World Meteorological Organization and the UN Environmental Programme to provide authoritative and up-to-date scientific information for policymakers. In 1990, the

9 COP21 and India's Intended Nationally Determined Contribution Mitigation Strategy   151

Table 9.1 Historical carbon space occupied by various countries in 2009 (1850 as base year)

Source	USA	Other developed countries	China	Other emerging economies	India
Share	29%	45%	10%	9%	3%

Source: Government of India Ministry of Environment, Forest and Climate Change (Government of India, 2010)

IPCC published its First Assessment Report confirming that climate change was indeed a threat and calling for a global treaty to address the problem. In May 1992, the Inter-Governmental Negotiating Committee adopted by consensus the formation of the United Nations Framework Convention on Climate Change (UNFCCC). To achieve its objectives, all parties who have ratified, accepted and approved the treaty are subject to an important set of general commitments. The supreme body of the convention is called the Conference of the Parties (COP) and its main task is to review the National Communications and Emission Inventories submitted by the parties. It organised the Earth Summit in Rio de Janeiro, Brazil, in June 1992. Governments of the world met at the Rio Earth Summit and agreed, among other things, to address climate change and pursue sustainable development. The concept of sustainable development was defined by the Brundtland Report (World Commission on Environment and Development 1987, p. 43, http://www.wsu.edu>susdev>WCED87) as "development that meets the needs of the present without compromising the ability of future generations to meet their own needs".

The UNFCCC divided countries into two groups: Annex I Parties and Annex II Parties. India is in Annex II. Subsequently, the IPCC published the Second Assessment Report in 1995, the Third Assessment Report in 2001, the Fourth Assessment Report in 2010 and the Fifth Assessment Report in 2015.

The first conference (COP1) was held in Berlin in 1995 and has been held regularly each year. COP8 was held in New Delhi in 2002 and COP21 was held in Paris in 2015.

Progress has been slow, as can be seen from the outcome of the different rounds of talks over the years:

- 1997—Kyoto Protocol: Annex 1 Parties commit to reduction targets.
- 2007—COP13 in Bali: Introduction of Nationally Appropriate Mitigation Actions, to engage developing countries in voluntary mitigation efforts.
- 2009–2010—COP15 in Copenhagen and COP16 in Cancun: Comprehensive International System for collective action and major developing countries announced mitigation pledges (including India).
- 2011—COP17 in Durban: Ad Hoc Working Group on the Durban Platform for enhanced Actions launched for evolving a new agreement for post-2020 period.
- 2013–2014—COP19 in Warsaw and COP20 in Lima: Intended Nationally Determined Contribution concept for all countries.
- 2013—COP19 in Warsaw: Developing countries have gradually assumed greater responsibilities. All countries required to prepare INDCs and present them before COP21 in Paris.

- 2014—COP20 in Lima: Further clarity that INDC is not mitigation centric and can include other components as per country priorities.

After more than two decades of disappointing progress, in 2015 at COP21, nations pledged to combat CO_2 emissions in the most effective way. The global agreement reached at COP21 was a decisive turning point for the world's efforts to fight climate change. COP21 achieved a legally binding and universal agreement on climate with the aim of keeping global warming below 2 °C. In the Paris Agreement, nations have agreed to undertake rapid reduction procedures to achieve a balance between anthropogenic emissions by sources and removal by sinks of greenhouse gases by 2050, at which time global greenhouse gas emissions must be reduced to zero if we are to limit warming to the 1.5–2 °C goal. COP21 gives us the last chance to make it possible to secure a safe climate (See UNFCCC, 2015).

9.1.3 The Achievements of the Paris Agreement at COP21

Two major programmes have been launched by the United Nations to address the problems of climate change and building sustainable economies: the United Nations Climatic Change Conference (the COP initiative) and the Sustainable Development Goals plan from 2016 to 2030 involving UN member nations. To strengthen the global response to climatic change with sustainable development was the main outcome of the 2015 COP21 Paris Agreement. Both the agendas have respective targets to be met by 2030. The Agreement commits the world's nations to take mitigation actions that will limit global warming to well below 2 °C above pre-industrial levels and pursue efforts to limit the temperature increase to 1.5 °C. As per the provisions of the Paris Agreement, the treaty has come into force as 55 countries, contributing 55% of total global emissions, have ratified the agreement.

The governments agreed on the following issues:

- A long-term goal of keeping the increase in global average temperature to well below 2 °C above pre-industrial levels.
- To aim to limit the increase to 1.5 °C, since this would significantly reduce the risk and impacts of climate change.
- The need for global emissions to peak as soon as possible, recognising that this will take longer for developing countries.
- To undertake rapid reductions thereafter, in accordance with the best available science.

In this respect, the INDC of each country specifies that nation's mitigation actions and targets. At 5-year intervals, a global stocktake will be undertaken to determine whether national mitigation actions in aggregate are on track to achieve the global warming 1.5–2 °C goal. If not, nations have agreed to increase the ambition of their mitigation commitments.

To help developing countries achieve their INDC the Green Climate Fund (GCF) was created.

Table 9.2 Key macro indicators reflecting India's future needs

Indicator	India in 2014	India in 2030
Population	1.2 (billion)	1.5 (billion)
Urban population	377 (million) in 2011	609 (million)
GDP at 2011–2012 prices (in trillion)	106.44 rupees (US$1.69)	397.35 rupees (US$6.31)
Per capita GDP in USD (Nominal)	1408	4205
Electricity demand (TWh)	776 (2012)	2499

Source: United Nations, Department of Economic and Social Affairs, Population Division (2014), Population Foundation of India (2016)

9.1.4 Green Climate Fund

The GCF is a financial mechanism within the UNFCCC and became operationalised at COP17 in Durban in 2011. The GCF was established by 194 governments to limit or reduce greenhouse gas emission. Its headquarters is in Songdo, South Korea, and it is governed by a board of 24 members and initially supported by a secretariat. The GCF has set itself a goal of raising US$100 billion per year by 2020. All developing country parties to the UNFCCC will get support for projects, programmes, policies and other activities from the GCF. They will receive the funding through accredited national, subnational and regional implementing entities and intermediaries.

9.2 India

9.2.1 Climate Change and Development

As the Indian population increases and its economy grows at a fast rate, there will be demographic change along with an increased demand for urbanisation. This will in turn increase the demand for housing, energy, transport, water and waste disposal. All these will have adverse effects on biocapacity.

Table 9.2 shows the projected key macro indicators reflecting India's future needs as the economy grows.

India ranked 5th in aggregate greenhouse gas emissions in the world, behind the USA, China, the EU and Russia. Interestingly, the emissions of the USA and China were almost four times that of India in 2007. According to a report by the World Resource Institute (2014), a global research organisation, India was the fourth-largest CO_2 emitter behind China, the USA and the EU. Thus, after ranking fifth in 2007, it became fourth in 2012.

Table 9.3 gives the World Resource Institute (2014) figures of percent of total emissions and ton of emissions per capita of different countries. It shows that India is far behind the top three emitters in terms of per capita emissions, with the USA having eight times India's emissions per capita. Developing countries China,

Table 9.3 Greenhouse gas emissions of different countries

Top 10 emitters		Top 10 per capita emitters	
Countries	Percent of total emission	Countries	Ton of emission per capita
China	25.26	USA	19.86
USA	14.4	Russia	16.22
EU	10.16	Japan	10.54
India	6.96	Iran	9.36
Russia	5.36	EU	8.77
Japan	3.11	China	8.13
Brazil	2.34	Mexico	5.99
Indonesia	1.76	Brazil	5.10
Mexico	1.67	Indonesia	3.08
Iran	1.65	India	2.44

Source: World Resource Institute (2014)

Mexico and Brazil also have higher per capita contributions compared to India. The top ten emitters contributed 72% of global greenhouse gas emissions (excluding land use, land use change and forestry (LULUCF)). On the other hand, the lowest 100 emitters contributed less than 3%.

The total net greenhouse gas (GHG) emissions from India in 2007 was 1727.71 million tons of CO_2 equivalent (eq), which is the total of CO_2, methane and nitrous oxide emitted in terms of their respective global warming potentials (Government of India, 2010). Of that 1727.71 tons, CO_2 emissions were 1221.76 million tons, methane emissions were 20.56 million tons and nitrous oxide emissions were 0.24 million tons.

India's per capita CO_2eq emissions, including LULUCF, were 1.5 tons/capita in 2007. India's population in 2007 was 1.15 billion, approximately 17% of the global population (The Energy and Resources Institute, 2011). The per capita greenhouse gas emission without LULUCF is estimated to be 1.7 tons of CO_2eq per capita and with LULUCF it is 1.5 tons/capita.

The total greenhouse gas emissions without LULUCF grew from 1251.95 million tons in 1994 to 1904.73 million tons in 2007 at a compounded annual growth rate of 3.3%; with LULUCF it was 2.9%. Between 1994 and 2007, some sectors indicated significant growth in greenhouse gas emissions, such as cement production (6.0%), electricity generation (5.6%) and transport (4.5%). Greenhouse gas emissions in 2007 from the energy, industry, agriculture and waste sectors constituted 58%, 22%, 17% and 3% of the net CO_2eq emissions, respectively.

Energy The energy sector emitted 1100.06 million tons of CO_2eq due to fossil fuel combustion in electricity generation, transport, commercial/institutional establishments, agriculture/fisheries and energy-intensive industries such as petroleum refining and manufacturing. Of that, 719.31 million tons of CO_2eq were emitted from electricity generation, including both grid and captive power. The CO_2eq emissions from electricity generation were 65.4% of the total CO_2eq emitted from the energy sector. Coal constituted about 90% of the total fuel mix used.

Petroleum Refining and Solid Fuel Manufacturing These energy-intensive industries emitted 33.85 million tons of CO_2eq in 2007 which was 3.1% of the total emissions. The solid fuels include manufacturing of coke and briquettes.

Transport Transport sector emissions are reported from road transport, aviation, railways and navigation. In 2007, 142.04 million tons of CO_2eq were emitted from the transport sector, which was 12.9% of the total emissions. Road transport, being the dominant mode of transport in the country, comprised 87% of the CO_2 equivalent emissions from the sector. The aviation sector, in comparison, only emitted 7% of the total CO_2eq emissions. The rest were emitted by the railways (5%) and navigation (1%) sectors. Bunker emissions from aviation and navigation have also been estimated but are not counted in the national totals.

Residential and Commercial The residential sector in India is one of the largest consumers of fuel outside the energy industries. The commercial/institutional sector uses oil and natural gas over and above conventional electricity for its power needs. In India, biomass constitutes the largest portion of the total fuel mix used. The total CO_2eq emissions from residential and commercial/institutional sectors was 139.51 million tons of CO_2eq in 2007, which was 12.6% of the total emission.

Agriculture and Fisheries The agriculture and fisheries activities together emitted 3.1% of the total emissions, or 33.66 million tons of CO_2eq, due to energy use in the sector other than grid electricity. Estimates of greenhouse gas emissions from the agriculture sector arise from enteric fermentation in livestock, manure management, rice paddy cultivation, and agricultural soils and on field burning of crop residue.

Industry The industry sector emitted 412.55 million tons of CO_2 eq.

Land Use, Land Use Change and Forestry The LULUCF sector is a net sink. It involves estimation of carbon stock changes, CO_2 emissions and removals and non-CO_2 greenhouse gas emissions. It sequestered 177.03 million tons of CO_2 in 2007. Estimates from the LULUCF sector include emissions by sources and or removal by sinks from changes in forest land, crop land, grassland and settlements.

9.2.2 India's Intended Nationally Determined Contribution (INDC) to Control Climate Change

On 28 September 2016, the President of India signed the Paris Agreement, and the Government of India formally ratified it on 2 October 2016, the birth anniversary of Mahatma Gandhi. India is committed to engaging actively in multilateral negotiations under the UNFCCC in a positive, creative and forward-looking manner. Keeping in mind its development agenda coupled with its commitment to follow the low-carbon path to progress, India communicated its INDC in response to COP decisions 1/CP.19 and 1/CP.20 for the period 2011 to 2030 (See UNFCCC, 2016):

1. To propagate healthy and sustainable way of living.
2. To adopt a climate-friendly and cleaner path of economic development.
3. To reduce CO_2 emission intensity of GDP by 35% of 2005 levels before 2030.
4. To achieve about 40% cumulative electric power installed capacity from non-fossil fuel-based energy resources by 2030, with the help of transfer of technology and low-cost international finance, including from Green Climate Fund.
5. To create an additional carbon sink of 2.5–3 billion tonnes of CO_2eq through additional forest and tree cover by 2030.
6. To better adapt to climate change by enhancing investment in development programmes in sectors vulnerable to climate change.
7. To mobilise domestic and additional funds for mitigation and adaptation actions.
8. To build capacities for cutting-edge R&D in this regard.

To achieve the above contributions, India is determined to continue with its ongoing interventions, enhancing existing policies and launching new programmes.

It is difficult to estimate the cost of achieving India's INDCs at this time. But the preliminary assessment indicates implementing climate mitigation and adaptation measures in India through INDCs from 2015 to 2030 will cost approximately US$2.5 trillion at 2014–2015 prices. India has developed INDCs on the basis of its development agenda with existing resources and capacity base.

To meet the funding requirements, India plans to mobilise both domestic and international resources; to get funding from international sources India is targeting the GCF.

The National Bank for Agriculture and Rural Development was fast-tracked as an accredited National Implementary Entity at the 10th board meeting of the GCF in July 2015. The bank is eligible to submit large projects having outlays of more than US$250 million. The Government of India is formulating its own rules for project identification and sanctions on the basis of criteria by which GCF has sanctioned eight projects (Government of India, 2015a). The bank is in talks with states and ministries for identification and appraisal of green projects, to see if they meet the requirements of the fund.

The major share of domestic funding of climate finance comes from different budgetary sources, as most of the resources for adaptation and mitigation are built into the ongoing sectoral programmes. A careful mix of market mechanisms, together with fiscal instruments and regulatory interventions, are made to augment the availability of funds.

India imposed a carbon tax called the Clean Energy Tax on 1 July 2010, but only on coal. It was initially set at 50 rupees (US$0.80) per ton of domestic and imported coal. In 2015, the tax quadrupled to 200 rupees (US$3.20) per ton of coal. The tax on coal production was renamed as the Clean Environment Tax, using the emission factor for coal of around US$2.00 per ton. So far, it has been a massive success for financing the National Clean Environment Fund used to invest in clean energy, technologies and related projects. The total collection of 170.84 billion Indian rupees (US$2.7 billion) up until 2014–2015 has been used for 46 clean energy projects worth 165.11 billion rupees (US$2.6 billion). In 2016, the coal tax was doubled from 200 rupees (US$3.20) to 400 rupees (US$6.40) per ton, to help finance more clean energy projects (Mondal, Irfan, & Ramaswamy, 2016).

India has also set up a National Adaptation Fund to address climate change adaptation needs with an initial allocation of 3500 million rupees (US$55.6million). This fund, in addition to sectoral spending by the respective ministries, will be used to fund sectors like agriculture, water, forestry, etc.

Subsidies Cut and Increased Taxes on Fossil Fuels (Petrol and Diesel) An important fiscal measure taken is a cut in subsidies and increased taxes on fossil fuels (petrol and diesel). This action leads to an implicit carbon tax (US$140 for petrol and US$64 for diesel) in absolute terms. This tax is considered to exceed the reasonable value in terms of an initial carbon tax and has led to a decrease in the petrol subsidy by approximately 25%. World Bank estimates suggest that these measures will help India to achieve a net reduction of 11 million tons of CO_2 emission in less than a year (Government of India, 2015b). In addition, India has implemented tax-free infrastructure bonds of 50 billion rupees (US$794 million) for funding of renewable energy projects during the year 2015–2016 (Government of India, 2015c).

Finance Commission Incentive for Creation of Carbon Sink This is the 14th Finance Commission recommendation on incentives for the forestry sector. This initiative has effectively given afforestation a massive boost, as US$6.9 billion will be transferred to the states. This will increase to US$12 billion by 2019–2020.

Renewable Energy Certificates Renewable energy certificates is a market instrument that endorses renewable energy and its participation in the energy market. This was established in 2010 by the Ministry of New and Renewable Energy (MNRE) and has been allowed to trade since March 2011.

Perform Achieve and Trade Perform Achieve and Trade aims to improve energy efficiency by 1–2% per year for energy-intensive industries. In eight energy-demanding sectors, 478 plants participated resulting in a 4–5% decline in energy consumption at the end of 2015 compared to 2012.

9.3 India's Mitigation Effort

According to the World Resource Institute report, the energy sector is the dominant source of greenhouse gas emissions, contributing 75% of global emissions. The International Energy Agency has estimated that absolute emissions in India have been increasing over the decades. The 2013 estimate stands at 1954.02 $MtCO_2e$ but the per capita emission of 1.7 tonnes of CO_2 falls significantly short of 4.4 tonnes of CO_2e. This trend is likely to continue until 2031. According to INCCA (2010, http://www.moef.nic.in/downloads/public-information/fin-rpt-incca.pdf), the primary emitting sectors in the country are electricity generation, transport, agriculture, industry, buildings, fugitive emissions and waste. Greenhouse gas emissions in 2007 from the energy sector constituted 58% of the net CO_2e emissions. The transport sector emissions are reported from road transport, aviation, railways and navigation. In 2007, 142.04 million tonnes of CO_2e were emitted from the transport sector which is 12.9% of the total emissions. It is noted that electricity and transport sectors together account for more than 70% of CO_2 emissions. Therefore, we have focused our discussion on the energy sector from a sustainable development and mitigation angle.

Energy Sector Non-commercial energy sources like firewood, dung cake and agricultural waste were the main sources of energy (more than 70%) of the total energy consumed in India at the time of independence. The situation rapidly changed with urbanisation and the changing demographic situation. In 1965, primary commercial energy consumption was 53 Million Tonnes of Oil Equivalent (MTOE) and per capita energy consumption was about 110 kg of oil equivalent. The primary energy consumption rose to 300 MTOE by 2000, which is six times that of 1965, and per capita energy consumption tripled to a little more than 300 kg of oil equivalent as the population doubled during this period.

Among the commercial energy sources, coal, due to its abundance, was a predominant source of energy, accounting for 55% (about 315 MTOE) of Indian energy consumption in 2002. This was followed by oil (31%), natural gas (8%), hydro (5%) and nuclear (1%) energy. Table 9.4 provides a breakdown over time.

As the GDP growth rate of India increases, the overall growth of the country shows the changing pattern of energy consumption. People are using electricity and gas as their primary source. The consumption of electricity is increasing with urbanisation.

Thermal Power In 2013, electricity coverage in India was only 81%; by 2050 it is estimated that there will be full coverage of electricity throughout India and per head consumption will also increase due to urbanisation. The Power Grid Corporation of India in 2013 estimated that the total demand for power is likely to reach 890,000 MW by 2050. The utility electricity sector in India has an installed capacity of 307.28 GW as of 31 October 2016. Of this, coal (thermal) plants produce 186,492.88 MW (60.7%), large-scale hydro 43,112.43 MW (14.0%), small-scale hydro 4323.33 MW (1.4%), wind power 28,082.95 MW (9.1%), biomass 4882.33 MW (1.6%), solar power 8513.23 MW (2.8%), gas 25,057.13 MW (8.2%), nuclear 5780 MW (1.9%) and diesel 918.89 MW (0.3%). Thus, renewable power plants constitute 28.9% of total installed capacity.

Coal is the most abundant source of commercial energy in India. The total coal reserves (as on 1 January 2014) have been assessed at about 301.56 billion tonnes.

Coal production increased rapidly after the nationalisation of the coal mines. From 72.9 million ton in 1970–1971, it increased to 211.7 million ton in 1990–1901 and to 327 million ton in 2001–2002, making India the world's fourth-largest coal producer. The increase is predominantly in non-coking coal production. One of the major constraints on the profitability of the coal sector is the low productivity levels in underground mines. These mines employ 80% of human resources, but contribute to only 30% of the total output of the sector. Since nationalisation of the industry, India's mine planners have chosen opencast mining over underground methods, to enhance productivity and meet production targets. The drawback of opencast methods is that its quality is unavoidably affected by contamination of overburden mixes into the coal.

9 COP21 and India's Intended Nationally Determined Contribution Mitigation Strategy 159

Table 9.4 India's primary energy consumption per capita

Year	Primary energy consumption (MTOE)	Oil (MTOE)	Gas (MTOE)	Coal (MTOE)	Nuclear (MTOE)	Hydroelectricity (MTOE)	Population (millions)	Energy consumption (per capita)
1965	53	13	0	36	0	4	485	109
1966	55	14	0	36	0	5	495	110
1967	56	15	0	36	0	5	506	111
1968	60	16	0	38	0	6	518	116
1969	67	20	0	40	0	7	529	126
1970	65	20	1	38	0	7	541	120
1971	67	21	1	38	0	8	554	121
1972	71	22	1	41	0	7	567	125
1973	72	23	1	40	0	8	580	124
1974	77	23	1	45	0	8	593	129
1975	82	23	1	49	1	9	607	135
1976	86	25	1	51	1	9	620	139
1977	91	26	1	53	1	10	634	143
1978	94	29	1	51	1	12	648	146
1979	99	31	1	54	1	12	664	150
1980	103	32	1	57	1	13	675	152
1981	113	34	2	63	1	13	690	163
1982	114	35	2	65	1	11	705	161
1983	120	37	2	69	1	11	720	167
1984	126	40	3	70	1	12	736	172
1985	137	43	4	77	1	12	751	182
1986	147	46	6	83	1	12	767	191
1987	155	47	6	90	1	11	784	198
1988	167	52	7	95	1	12	801	208

(continued)

Table 9.4 (continued)

Year	Primary energy consumption (MTOE)	Oil (MTOE)	Gas (MTOE)	Coal (MTOE)	Nuclear (MTOE)	Hydroelectricity (MTOE)	Population (millions)	Energy consumption (per capita)
1989	185	56	10	104	1	14	817	226
1990	193	58	11	108	1	15	835	232
1991	206	59	13	116	1	17	852	242
1992	217	62	14	123	1	16	868	250
1993	223	63	15	128	1	16	886	251
1994	236	67	16	134	1	18	904	261
1995	252	73	18	143	2	17	922	274
1996	270	79	19	154	2	16	940	287
1997	281	83	19	160	2	16	955	294
1998	290	87	22	160	3	19	971	299
1999	297	95	22	158	3	19	987	301
2000	313	98	23	171	4	17	1002	313
2001	315	97	24	174	4	16	1027	306

Source: Reliance Industries Limited (2002)

There are 132 thermal plants with 431 units in India. The Centre for Study for Science and Environment estimates that coal-based thermal plants are the least efficient and therefore the most polluting in the world (The Centre for Study for Science and Environment, 2015). Currently, India's electricity generation division is accountable for 38% of the total greenhouse gas emissions. It is projected that more than 3000 million tonnes of CO_2 will be released by 2030 as India's use of coal is projected to triple by 2030 as energy demand doubles due to economic growth and rapid industrialisation coupled with accelerated urbanisation.

In order to meet the demand and at the same time reduce emissions and make the thermal plants units efficient, the Government of India has initiated a number of actions. It has been estimated that during the Twelfth Plan (2012–2017) there will be an additional 50% coal-based capacity and in the Thirteenth Plan (2017–2022) there will be 100% coal-based capacity additions through new technologies called ultra-super critical technology (Government of India, 2013). These technologies are 10% more efficient and reduce CO_2 emissions by 50%. Similarly, coal-based methane technology will be developed during the next plan period. The transmission loss that accounts for 25% of energy will be mitigated through using new transmission technology.

Renewable Energy In 1970, India formulated its renewable energy programme in response to the development of clean technology and alternative sources of energy to satisfy energy demand and develop a climate change mitigation strategy. The Commission for Additional Sources of Energy was created in 1981 to exploit the possibility of getting renewable energy from the inexhaustible supply of sunlight, wind and biomass. In 1987, the Indian Renewable Energy Department Agency was created to find renewable energy initiatives. In 2006, the commission was transformed into the MNRE and, through it, funding of renewable energy initiatives and policy formulations began.

The primary thrust in the renewable sector in the country came in with the Electricity Act 2003. It has accelerated and motivated India's search for alternative energy sources to fossil fuels while replacing all prior legislation related to India's electricity. It introduced crucial policy instruments like the National Electricity Policy, a tariff policy, and the National Electricity Plan that optimises the use of resources including thermal, nuclear, hydro and mainly renewable. The National Electricity Policy encourages the use of renewable energy by directing the State Electricity Regulatory Commissions to launch a tariff for electricity generated from renewable sources to be competitive in the energy market. It also stresses the importance of using renewable energy for preserving and sustaining the environment and mitigation of climate changes.

India is blessed with an immense potential of renewable energy resources to meet its energy demands. Table 9.5 gives the cumulative capacity up to March 2012.

Solar Energy India's potential solar energy is 5000 trillion kWh per year. It is located between Tropic of Cancer and the equator with average temperatures of 25–27 °C, and it receives solar power of 4–7 kWh solar radiation per square metre for 250–300 sunny days annually. Initially, solar power was not considered to play a significant role because of high costs and lack of technological know-how. India was using solar energy earlier in a reduced capacity but with the launch of the Jawaharlal Nehru National Solar Mission, solar power received greater attention. This mission has set an ambitious programme of deploying 20,000 MW of grid-connected solar power by 2022. The mission also had the objective of reducing the cost of solar power generation in the country, large-scale deployment goals, aggressive R&D and the domestic production of critical raw materials, components and products.

The MNRE has set a target of updating solar energy to 100 GW by 2022. India has rapidly increased its solar power capacity from 3.7 MW in 2005 to 4060 MW in 2015 with a compound annual growth rate exceeding 100% over the past 10 years. The solar sector has grown more than 200-fold in the last 5 years and is expected to expand further under the solar mission to install 22 GW of solar capacity by 2022. As of 30 April 2016, solar power was below 7% of its ambitious target although expansion is expected to be dramatic in the near future.

Wind Energy Wind energy is the largest growing sector in the renewable energy generation programme in India. Wind power accounts for 14% of India's total installed power capacity. The potential of wind firms in the country has been assessed as 302,000 MW at 100 m hub height. The wind resource at higher hub heights that are prevailing is possibly even more.

Wind energy production began in 1986 with the setting up of wind firms in the coastal areas of India by the MNRE, and its capacity has significantly increased in the last few years. In 2002, the installed capacity was only 1667 MW but it rose to 23,762 MW by 31 March 2015. As of 31 August 2016, the installed capacity of wind power in India had become 27,676 MW. Unfortunately, wind energy distribution is not uniform; it is mainly spread across the south, west and north regions, with the east and north-east regions having no grid-connected wind power plants as of March 2015.

India's wind industry is growing 2.1 GW a year by additional installations and is projected to save 48 million tons of CO_2 in 2020 and 105 million tons in 2030. India's current target is to attain 60 GW of wind power installation capacity by 2022.

Biomass Biomass is the term used to describe all plant-derived materials. It may be used to generate energy by direct combustion or by conversion to either a liquid or a gaseous fuel. The world derives a little over 10% of its energy from biomass. It is an important energy source in China and India, accounting for 20% and 40%, respectively, of the primary energy supply. About 70% of India's population relies on biomass because of the benefits it offers: it is renewable, widely available and carbon-neutral, and it has the potential to provide significant employment in rural areas. Biomass energy technology is also rapidly advancing. Besides direct combustion, techniques for gasification, fermentation and anaerobic digestion are all increasing the potential of biomass as a sustainable energy source.

Biomass energy accounts for 13% of total renewable energy generation in India. India's overall biomass potential has been assessed at about 19,500 MW, inclusive

Table 9.5 Potential of green new renewable power potential GW

Technology	Potential GW	Capacity factor	Generation bill KWh	Cumulative capacity GW March 2012
SPV	850	0.2	1489.2	1.579605
SPV Pump & Panel Over banks & canal	322	0.2	564.144	
Wind on shore	2006	0.25	4393.14	15.864622
Wind off shore	15	0.25	32.85	
Biomass	18	0.6	94.608	3.99187
Cogeneration from bagasse	5	0.6	26.28	
Waste to energy	7	0.6	36.792	0.078368
Other Sources: Geo Thermal and Ocean	16	0.2	28.032	
Small Hydro < 25 MW	15	0.2	26.28	2.975535
TOTAL New renewable	3254	–	6691.326	24.49
Large Hydro > 25 MW	150	0.2	262.8	38.99
Total green power	3404	–	6954.126	63.48

Source: Ernst and Young (2013)

of surplus power from bagasse-based cogeneration. India is the world's largest producer of sugar cane, and a potential of 3500 MW of power generation through bagasse-based cogeneration in sugar mills has been identified. Biomass power generation in India is an industry that attracts investment of more than 60 million rupees every year, with yearly employment of more than 10 million human-days in rural areas. The MNRE has been supporting the promotion of the biomass energy programme since the mid-1999s through favourable regimes at state and central levels. India aims to increase its biomass installed capacity to 9 GW by 2022 through its biomass power and cogeneration programme. The MNRE is persuading state governments and state electricity boards to announce remunerative policies for purchase, wheeling and banking of power generated from biomass projects.

Hydropower Hydroelectricity is the most important source of renewable energy and the second largest source of energy next to thermal power. A (United Nations Development Programme) study puts the world's gross theoretical potential of hydropower at about 40,500 TWh, technical potential at about 14,320 TWh and economic potential at about 8100 TWh. About one-third of this has so far been developed worldwide. Although hydroelectricity is generally considered a clean energy source, it is not totally devoid of greenhouse gas emissions, ecosystem burdens and socioeconomic adversities. Recently, large-scale hydropower projects have faced considerable environmental and social pressures and closed due to financial constraints. The share of hydropower generation has gradually declined in India over the last 30 years.

As of 2013, India was ranked fifth in available hydropotential globally, and hydropower provides 17% of the total electricity generated in India. Hydropower has a growing potential of more than 150 GW. Currently it has an installed capacity of 4 GW and an estimated potential of 19.7 GW. India aims to increase the installed small hydropower capacity to 10,000 MW by 2022. The MNRE provides incentives to encourage projects by subsidising 75% of the costs, and some states have deployed special tariffs, low-interest loans and other financial mechanisms to encourage private sector entrepreneur developments and investments. A number of programmes and projects are advocating new and efficient schemes for watermill designs which are effective in remote and rural areas because they can be deployed on small rivers, canals and dams.

9.4 Impediments in Meeting the INDC

Some impediments are coming up in trying to meet the commitments of the INDC. These arise due to the national and international scenario, viz. demonetisation of the Indian currency, the USA's withdrawal from Paris commitments and increase in world oil prices.

9.4.1 Demonetisation

The Interim Report of the Expert Group on Low Carbon Strategy for Inclusive Growth (2011) indicated that a 25% reduction in emission intensity could translate into emissions of to 3537 $MtCO_2eq$ in 2020 compared to 1433 $MtCO_2eq$ in 2005, assuming a GDP growth rate of 8% per year (Planning Commission of India, 2014). On 8 November 2016 at 8 PM, India's Prime Minister suddenly announced that from midnight the high-value 500- and 1000-rupee notes would no longer be valid. He gave citizens 50 days to deposit these notes into the banks. He said that these decisions were to clean black market money, reduce the risk of terrorism and restrict withdrawal of money to new notes.

It was not a well thought-out decision. Dr. Manmohan Singh, former Indian Prime Minister and renowned economist, predicted that GDP growth would fall by 2%. The growth rate of the country for the last few years has been around 7.5%; if it declines by 2% then it would be difficult to reduce emission intensity by 25%.

9.4.2 Oil Price

The Organization of Petroleum Exporting Countries decided to cut oil production by 1.2 million barrels per day from 1 January 2017. Other countries also decided to cut production by 0.6 million barrels per day from the same date. As a result, in the international market, oil prices rose from US$35 to US$50 per barrel and may reach US$60.

India enjoyed the benefits of falling crude prices in the international market from 2014 to 2015 as it helped control the inflation rate. Green projects may face a fiscal imbalance as crude oil prices jump sharply. In India, oil demand is projected to rise to more than 10 million barrels per day by 2040 compared to 4.1 million in 2015.

9.4.3 US Policy

President Donald Trump has announced that the USA is withdrawing from the Paris Accord, which has far-reaching consequences in world and Indian environment policy.

The commitment of US$100 billion from now until 2020 for the GCF will be difficult to collect and will affect India's aim to get funds for its green projects from GCF.

9.5 Conclusion

The emission intensity of India's GDP declined by more than 30% during the period 1994–2007, due to the efforts and policies that India was following. India wants to further reduce the emission intensity of GDP by 20–25% between 2005 and 2020 following the path of inclusive growth.

But the surprise demonetisation by the Prime Minister of India of the high-value notes has caused total imbalance of the economy. Experts say that GDP growth rate will fall by 2% and the inflation rate will also be high. In this situation, it will be difficult for the country to fulfil its commitments.

The global scenario is also changing and will have great impact on the funding of the INDC. If the funding of projects becomes uncertain, their implementation will be difficult.

Let us hope that in the long run the effect of demonetisation will be good and the Indian economy recovers and can fulfil its commitments.

References

Centre for Study for Science. (2015). *India's first-ever environmental rating of coal-based power plants finds the sector's performance to be way below global benchmarks*. Retrieved from http://www.cseindia.org/content/india%E2%80%99s-first-ever-environmental-rating-coal-based-power-plants-finds-sector%E2%80%99s-performance

Ernst and Young. (2013). *Mapping India's renewable energy growth potential: Status and outlook 2013*. Retrieved from http://www.ey.com/Publication/vwLUAssets/Mapping_Indias_Renewable_Energy_growth_potential/$FILE/EY-Mapping-Indias-Renewable-Energy-growth-potential.pdf

Government of India. (2010). *India: Greenhouse gas emissions 2007 Ministry of Environment and Forest Govt. of India*. Retrieved from http://www.moef.nic.in/downloads/public-information/Report_INCCA.pdf

Government of India. (2013). *Twelfth five year plan (2012–2017) Economic sectors*. Retrieved from http://planningcommission.gov.in/plans/planrel/12thplan/pdf/12fyp_vol2.pdf

Government of India. (2015a). Retrieved from http://envfor.nic.in/tenders-advertisements/call-proposal-national-implementing-entitiesnies-green-climate-fund-set-under

Government of India. (2015b). *Financial support for renewable energy press release Ministry of Overseas Indian Affairs 30-November-2015*. Retrieved from http://pib.nic.in/newsite/PrintRelease.aspx?relid=132129

Government of India. (2015c). From carbon subsidy to carbon tax: India's green actions, Chapter 9. *India Budget, 2014–2015*. Retrieved January 10, 2017 from http://indiabudget.nic.in/es2014-15/echapvol1-09.pdf

INCCA. (2010). *Climate change and India: a 4 x4 assessment a sectoral and regional analysis for 2030s*. New Delhi, India: Government of India.

IPCC. (2014). Climate change (2014) synthesis report. In R. K. Pachauri & L. A. Meyer (Eds.), *Contribution of Working Groups I, II and III to the Fifth Assessment Report of the Intergovernmental Panel on Climate Change*. Geneva, Switzerland: IPCC.

Mondal, M., Irfan, Z. B., & Ramaswamy, S. (2016). *Does carbon tax makes sense? Assessing global scenario and addressing indian perspective* (N6. 2016-141).

Planning Commission of India. (2014). *The final report of The Expert Group on Low Carbon Strategies for Inclusive Growth*. Retrieved January 9, 2017 from https://www.google.com.au/search?q=Interim+Report+of+the+Expert+Group+on+Low+Carbon+Strategy+for+Inclusive+Growth&rlz=1C1GGRV_enAU751AU752&oq=Interim+Report+of+the+Expert+Group+on+Low+Carbon+Strategy+for+Inclusive+Growth&aqs=chrome..69i57.286j0j7&sourceid=chrome&ie=UTF-8

Population Foundation of India. (2016). *Annual report 2015–2016*. Retrieved from http://population-foundation.in/files/fileattached/Fileattached-1480318575-ANNUAL-REPORT-2015-2016.pdf.

Reliance Industries Limited. (2002). *Reliance review of Energy Markets Energy Research Group*. Retrieved from http://www.ril.com/getattachment/4c897441-74a4-4203-9a89-1c869139b676/AnnualReport_2002-03.aspx

The Energy and Resources Institute. (2011) *TERI Energy Data Directory & Yearbook (TEDDY) 2010*. New Delhi, India: TERI press.

The Footprint Network. (2016). *We would need 1.7 Earths to make our consumption sustainable*. Retrieved February 16, 2017 from www.footprintnetwork.org

UNFCCC. (2007). *Vulnerability and adaptation to climate change in small island developing States— Background paper for the expert meeting on adaptation for small island developing states*. Bonn, Germany: UNFCCC Secretariat. Retrieved 2016, from http://unfccc.int/files/adaptation/adverse_effects_and_response_measures_art_48/application/pdf/200702_sids_adaptation_bg.pdf

UNFCCC. (2008). *Report of the conference of the parties on its thirteenth session, from conference of parties 13, Bali, 3rd–15th December, 2007*.

UNFCCC. (2015). *Paris agreement*. Paris, France: United Nations.

UNFCCC. (2016) *India's nationally determined contribution*. Retrieved from http://www4.unfccc.int/ndcregistry/PublishedDocuments/India%20First/INDIA%20INDC%20TO%20UNFCCC.pdf

United Nations. Department of Economic and Social Affairs, Population Division. (2014). *World urbanization prospects: The 2014 revision, highlights* (ST/ESA/SER.A/352).

World Commission on Environment and Development. (1987) *The Brundtland Report*. New York: United Nations.

World Resources Institute. (2014). *The history of carbon dioxide emissions*. Retrieved from http://www.wri.org/blog/2014/05/history-carbon-dioxide-emissions

Chapter 10
Seasonal Drought Thresholds and Internal Migration for Adaptation: Lessons from Northern Bangladesh

Mohammad Ehsanul Kabir, Peter Davey, Silvia Serrao-Neumann, and Moazzem Hossain

Abstract It is widely accepted that human mobility caused by environmental change will take place internally within the affected countries rather than across borders. This research examines the link between environmental vulnerability and human migration in various socioeconomic contexts. Previous studies have examined population mobility in response to vulnerability driven by sudden natural hazards like cyclone, flood and earthquake. However, little is known about the dynamics of human mobility in response to slow-onset hazards like drought. This study included comprehensive fieldwork with socioeconomically disadvantaged migrants in northern rural areas of Bangladesh who are exposed to seasonal drought. The study focused on a better understanding of how affected individuals and families make decisions to either stay or to migrate internally in response to seasonal drought and other socioeconomic vulnerabilities. By adopting a case study approach, rural-to-urban migrants and their family members in the northern highland area of the country, known as the Barind Tract, were interviewed. The results suggest that migration decisions are consolidated by a variety of stressors including both environmental and nonenvironmental components. The research found that some interventions implemented by the government and nongovernment organisations are posing long-lasting impacts on the sustainability of rural livelihood with a propensity to increase, not reduce, outward migration. These interventions have been questions and recommendations are made to address this emerging and complex livelihood problem.

M.E. Kabir (✉) • P. Davey • S. Serrao-Neumann
Griffith School of Environment, Griffith University, Nathan, QLD, Australia
e-mail: ehsanul.kabir@griffithuni.edu.au; peter.davey@griffith.edu.au; s.serrao-neumann@griffith.edu.au

M. Hossain
Griffith Asia Institute and Department of International Business and Asian Studies, Griffith University, Nathan, QLD, Australia
e-mail: m.hossain@griffith.edu.au

© Springer International Publishing AG 2018
M. Hossain et al. (eds.), *Pathways to a Sustainable Economy*,
DOI 10.1007/978-3-319-67702-6_10

Keywords Migration • Internal displacement • Drought • Vulnerability • Threshold • Livelihood • Sustainability • Climate change • Microcredit and adaptation

10.1 Introduction

Moving forward from the discussions and outcomes of the United Nations Conference on Sustainable Development (Rio+20), internal migration is being recognised for its increasing relevance to social, economic and environmental dimensions of sustainable development, especially regarding livelihood adaptation options in response to environmental and climate change. Recent studies (including Gemenne, 2013; IOM, 2010a) suggest that migration is more of a survival strategy for the marginal poor during environmental crises, involving both sudden and slow-onset environmental hazards. Migration can help combat poor livelihood challenges and help households and communities cope with adverse environmental effects. Moreover, these migrants generate new workplace skills. However, it was discovered that remittances of monies back to families at their place of origin increase the potential to restore local livelihoods that have been severely distressed by various social and environmental challenges (IOM, 2010b; UNDESA–DSD, 2011). A new approach to planned migration is a key factor for the future of rural livelihood sustainability, which promotes economic growth and poverty reduction in origin locations.

While a large number of studies report increased human migration flows due to climate change and other environmental reasons, the empirical evidence explaining the multifaceted causes for this phenomenon is inconclusive (Koubi, Spilker, Schaffer, & Bernauer, 2016). A number of studies have reviewed the impacts of migration on the environment from excessive land use, soil degradation and social conflicts (e.g. Biermann & Boas, 2010 ; IOM, 2009 ; Reuveny, 2007 ; Vrij, Van der Steen, & Koppelaar, 1994). Despite widespread pressure from the media and mostly "grey literature", the claim suffers from a lack of evidence; more examination is required of how environmental factors affect outmigration, especially in the context of rural areas (IOM, 2010b). This study attempts to bridge the gap in the empirical evidence about migration by analysing the microlevel (i.e. household) decision-making of internal migrants from rural areas.

First, the research examined how different drivers (and their components) of vulnerability, including environmental hazards, influence internal migration decisions. This research is focused on seasonal drought in northern Bangladesh, described as a slow-onset environmental event. Second, the study participants identified a critical driver that impacted their decision to migrate. A more in-depth explanation of the threshold of rural–urban migration was sought, referred to in this study as the "breaking point" decision factor that invokes migration (Bardsley and Hugo, 2010). In support of this major finding, secondary data on diminishing groundwater levels was gathered to gain a clearer understanding of water insecurity and its impact on migration over the last 15 years. However, the literature suggests that generalisation of threshold points is not viable, as migration is highly contextual (Khandker, 2012; Mallick & Vogt, 2014; Qin, 2010). Nevertheless, it is still important to understand the threshold points of migration, to effectively plan and develop appropriate policy interventions.

The purpose of this study is to understand under what circumstances disadvantaged individuals living in the seasonal drought-prone area of Bangladesh decide to migrate internally. Migration is an adaptation to various environmental and nonenvironmental drivers of vulnerability. The literature does not provide any explanations for drought-affected internal migration. The fieldwork was conducted in the northern highland area of Bangladesh, known as the Barind Tract. Due to its height above sea level, the Barind region is largely vulnerable to seasonal drought and water scarcity for household consumption and agricultural irrigation. This study has focused on internal migrants from this drought-impacted region.

The study is significant for a number of reasons. First, among countries with a population greater than 10 million, Bangladesh is the most densely populated, having 1048.3 people per square kilometre and with a population growth rate of 1.2% (UN, 2015). Second, the country is dependent on natural resources, especially on land; the agriculture sector employs about 45% of the total labour force (CIA, 2011). Finally, in the context of Bangladesh, where insufficient agricultural land and shortfall of irrigation facilities in the northern Barind region are highly critical factors embedded in the livelihoods of the rural people, environmental hazards, including seasonal drought, could be considered as a possible key driver for rural–urban migration (Mallick & Vogt, 2014).

10.2 Literature Review

10.2.1 Internal Migration and Economic Well-being in Developing Countries

Several longitudinal studies claim that most internal migration brings positive economic changes for a nation. For instance, a study conducted by Beegle, de Weerdt, and Dercon (2011) from 1991 to 2004, tracking internal migrants leaving villages in the Kagera District of Tanzania, shows significant growth in migrants' family members' monthly consumption compared to people in nonmigrant families. A similar study by de Brauw, Mueller, and Woldehanna (2013) tracked migrants and nonmigrants from 18 villages in Ethiopia between 1994 and 2009. The results show that households of both rural-to-rural and rural-to-urban migrants from these villages have significantly higher consumption levels compared with nonmigrants (de Brauw et al., 2013). Although remittances of monies from newly settled internal migrants back to family members at their place of origin are much lower compared to similar remittances sent from overseas, these transfers significantly help families recover from poverty from lost income, debt and livelihood uncertainties (Adams, Cuecuecha, & Page, 2008; Wouterse, 2010). Hence, in the developing countries' context, it is often reported that households at the place of origin are better off when family members have migrated to cities to work (Ackah & Medvedev, 2012).

In low- or low–middle-income countries in South Asia, like Bangladesh, Pakistan and India, internal migration is often seen as the last resort, when no options are left to adjust to poverty and environmental crises at home (Mallick & Vogt, 2014). At the microlevel, the rural livelihood of these densely populated countries is constrained by a range of factors. Often, temporary migration of a family member can aid recovery from a crisis in the short term, by enhancing income and so reducing poverty (IOM, 2010a). In brief, where larger rural populations are vulnerable to environmental changes and socioeconomic difficulties, internal migration for employment is becoming an increasingly popular alternative livelihood and an adaptation option for many drought-affected communities.

10.2.2 Linking Environmental Changes with Migration: Evidence

The dynamics of human mobility in response to natural hazards is complex. One study investigating the effect of rainfall on migration of agricultural households in rural areas of Mali shows that although outmigration is theoretically expected to increase during drought, only the proportion of women out-migrating has increased slightly, keeping the total (not many men are migrating) population almost the same (Findley, 1994). Ahsan, Karuppannan, and Kellett (2011) found that rural–urban migration significantly increased in low-income countries like Bangladesh following drought and other natural hazards (see also Gray & Mueller, 2012; Kartiki, 2011). Additionally, internal migration in Bangladesh is characterised as mainly seasonal, due to poor rural economies; in many cases, natural hazards are just one of the major driving forces (see Afsar, 2000, 2003; Khandker, 2012; Qin, 2010). Nonetheless, loss of agricultural land and land salinity intrusion due to salt water shrimp farming in some coastal area of Bangladesh, classified as man-made environmental degradation, can compound the driving forces to migrate (Kabir, 2013).

Current studies in Bangladesh (including Ahsan et al., 2011; Mallick & Vogt, 2014) have found a positive relationship between sudden hazards like cyclone or flood and internal migration. However, in a longitudinal study conducted by Gray and Mueller (2012), the authors claimed that the consequence of flooding on long-term population displacement is "controversial" and that such a relationship cannot be strengthened through cross-sectional or quantitative studies only. In contrast, as argued by Gray and Mueller (2012), crop failure not related to flooding (i.e. caused by drought) in the northern Bangladesh may have greater influence on migration decisions at individual and household levels. This longitudinal study also indicates that individual farmers and agricultural labourers who are dependent on farming are more likely to migrate during seasonal drought. Such a noted exception clearly indicates the need for an alternative paradigm for framing research on environmentally influenced migration in the context of low-income countries like Bangladesh. Future studies should also consider components of vulnerabilities involving impacts and barriers on individuals and groups to migrate, in contrast with opportunities to stay by enhancing local adaptive capacity (Gray & Muller, 2012). Hence, there is a gap and uncertainty in the previous research findings. This qualitative investigation reviews the causal linkages between seasonal drought and vulnerable people and their decision-making approaches to migration internally in Bangladesh.

10.2.3 Drivers of Vulnerability to Explain the Migration Threshold

The theoretical basis linking vulnerabilities of individuals and groups/families and their decisions on migration requires investigation. Bardsley and Hugo (2010) define the "threshold" of migration as the exact circumstance that would make individuals or households take the final decision to move out (Bardsley and Hugo, 2010). This research endeavours to better understand the threshold of migration, which is in fact a well-known concept. According to Wolpert's original "Stress Threshold Model" (Wolpert, 1966), migration was seen as a response to the stress that residents are experiencing at a particular place (Wolpert, 1966). Wolpert introduced the term residential "stressors" that may increase the individual's "strain" level and drive residents to search alternative livelihood and dwelling places (Hagen-Zanker, 2008; Wolpert, 1966). Residential stressors in this study include pollution, environmental degradation and congestion. Speare (1974) further developed this model by emphasising that individual characteristics, social bond and location of residential houses are important interacting, compounding factors, and that together with environmental "stressors" determine the ultimate "strain". Thus, location-specific strains become key factors in involuntary migration.

As Bardsley and Hugo, (2010) further explained, the migration threshold, or breaking point, is always context-specific. Hence, attempts need to be made to illustrate a community's or individual's level of tolerance to stressors like extreme environmental situations. This is because environmental stress will have multiple interactions within the fabric of society (including economic, governance and cultural factors), upon which the highly contextual threshold for individuals will be determined. Individuals may take proactive diversification strategies with help from organisations at the institutional and grassroots levels to cope with the stressors (Boano, Zetter, & Morris, 2007; Kniveton, Schmidt-Verkerk, Smith, & Black, 2008). At the individual level, the threshold coping strategies can be specific to individual characteristics including skills, wealth, education, experience and social networks and how these traits dictate a response to certain adverse situations, like natural hazards (Murray et al., 2012). However, an effective example of how threshold levels are reached is still understudied. As Kniveton et al. (2008) righty argued, more information should be collected on how individuals make the decision to stay or delay their timing of migration in response to current stresses and coping strategy; as yet such explanations are scarce. Future studies examining the migration threshold need to be more illustrative to bridge policy and implementation gaps in support of disadvantaged migrants (Bardsley and Hugo, 2010).

One pragmatic shortcoming in assessing migration threshold is that rigorous methodological steps are not available to validate the concept with empirical data. The reason is that the migration threshold is seen as incorporating complex interactions between socioeconomic and biophysical dimensions (Bardsley and Hugo, 2010), and it is difficult to find empirical evidence for validation of the concept. One initiative to bridge the gap would be to measure it in other ways; however, methodological innovations are needed. In cases where primary data collection is concerned, the widely accepted qualitative and quantitative methods in hazard research and proxy data can be used to validate the concept. When the potential for harm from a diverse range of hazards is concerned, the *vulnerabilities assessment* is the widely accepted approach used for policy decision-making (Brooks, Adger, & Kelly, 2005; Tate, 2012). Vulnerability is described as the propensity of individuals or groups to be adversely affected by a system (Fekete, 2009; Michalos et al., 2011). This definition encompasses the idea of exposure to various stressors driven by harmful environmental, social, economic, political and institutional components. The examination of vulnerability and its various drivers and components would offer the flexibility to simplify and distil complex, real-world information, breaking down the layers of thresholds into a format useful for policymakers (Dickin, Schuster-Wallace, & Elliott, 2013).

10.3 Methodology

10.3.1 Study Location: Tanore Sub-district in Bangladesh

This research has followed a case study approach focusing on the Tanore Sub-district of the Rajshahi District located in the northern high Barind area. The Tanore Sub-district has been ranked as the second-most vulnerable area in Bangladesh due to crop failure linked with seasonal drought, as reported by the Bangladesh Ministry of Disaster Management and Relief (CDMP, 2013). The physiographical features of Tanore incorporate the highland Barind Tract (81.8%), the River Tista Floodplain (4.8%), the River Old Gangetic Floodplain (3%) and others including homesteads, wetlands, ponds and river (10.4%) (SRDI, 2000). One unique feature of this site is that the Rajshahi District has high rates of rural-to-urban migration (41% of total population mobility in the last 5 years) compared to other districts in Bangladesh (BBS, 2011). It cannot be assumed that such a high outmigration ratio could be linked to seasonal drought impacts on the livelihood of marginal farmers and other disadvantaged groups living at Tanore; more research is required.

10.3.2 Methods for Data Collection and Sampling

Primary data were obtained as part of the PhD doctoral fieldwork (from February to April 2017) of the first author of this chapter in the Tanore Sub-district. The local government administers each of the sub-districts in Bangladesh, which are divided into several unions.[1] Tanore Sub-district is a combination of seven unions and two municipalities (only in urban areas). First, the local government representatives, nongovernment organisation (NGO) representatives and community key informants were consulted to prepare a list of the unions and villages most adversely affected by seasonal drought within the Tanore Sub-district. Based on the key informants' advice, three unions and one municipality were selected for the research. The researcher visited villages in each of the shortlisted unions and municipalities to explore the individual cases of outmigration.

Overall, 26 cases of outmigration were traced and the migrant (the person who migrated leaving family at home, hence only occasionally found at home) or a family member (mainly wife or parent) present at each of the selected houses were interviewed. Two focus group discussions were conducted with the short-term migrants. In addition, six key informant interviews, involving the elected local government representative, official community key informants, a local NGO representative and a senior government geohydrologist, have enriched the study.

The purposive sampling[2] technique was adopted to identify households from where individuals migrated. Hence, the research partly relied on snowballing[3] (or chain-referral) sampling to increase the possibility of finding more cases of outmigration for the study. Once a potential household of a migrant was identified, an adult was approached for an interview. The participants were also asked whether they knew other households of outmigrants within the same village or community. Thus, with the help of some of the respondents, other cases of outmigration were traced.

Secondary data were collected from government records of groundwater levels from the Bangladesh Water Development Board which reports to the Ministry of Water Resources. Groundwater levels from six sites in the Tanore Sub-district between 2000 and 2014 were obtained.

---

[1] The area within each sub-district (except for those in metropolitan areas) is divided into several unions. Each union comprises multiple villages. Direct elections are held for each union to choose local government representatives, as per the nation's *Local Government Act*, No. 20, 1997.

[2] In sociology and statistics research, purposive sampling is a nonprobability sampling technique where the researcher relies on his or her own judgement when choosing or recruiting participants in the study.

[3] In sociology and statistics research, snowball sampling or chain-referral sampling is a nonprobability sampling technique where existing study respondents identify or recruit future respondents from among their contacts.

10.3.3 Listing of the Critical Drivers of Vulnerability

During the 10 weeks of interviewing outmigrants in the seasonal drought-affected areas in Tanore Sub-district, 26 cases (7 female and 19 male) were documented. All interviews involved a structured questionnaire with face-to-face discussions consisting of both closed- and open-ended questions. Methodologically, the open-ended questions provided an opportunity to discuss various drivers of vulnerability, involving socioeconomic and environmental components affecting the livelihood of the studied rural families. The respondents were asked to describe various drivers of vulnerability (i.e. strains/stressors) and the reasons influencing their decision to internally migrate. The majority of the respondents perceived the drivers of financial vulnerability as poverty based on no money to buy land for agriculture and no money for increasing farming expenses.

A list of drivers of vulnerability has been identified from an analysis of the interviews. The study participants were encouraged to select a range of environmental drivers, including water logging, seasonal drought, water scarcity and impact from storms as well as also nonenvironmental drivers of vulnerability, including poverty, death of a family member who was a primary earner, chronic illness, accident and microcredit burden among others. When participants were not able to relate to more than two drivers of vulnerability during their initial conversation with the researcher, a probing or prompting technique[4] was used to verify the presence of some other stressors in their livelihood. For example, if a respondent could only perceive poverty and water crises as the key drivers of vulnerability, they were prompted and asked again whether they were also exposed to other stressors like water logging, hail storm or a serious medical condition within the last 5 years.

A list of such drivers of vulnerability were documented during the field studies and continuously updated throughout the research. Finally, each respondent was asked to consider their critical driver or self-identified driver—the individual's breaking point or the reason that the final decision to migrate was reached to search for an alternative livelihood.

The critical drivers are listed below and explain the threshold points of migration discovered in this research:

1. Lack of income—lack of consistent yearly income caused by decreasing or no income or opportunity in the study area.
2. Seasonal lack of income—relevant to people who can generally find short-termed local employment opportunities at different times of the year, such as seasonal agricultural labourers and freshwater fishermen who alternate jobs locally between rainy and harvesting seasons.
3. Loss in agriculture—environmental and nonenvironmental reasons where the agricultural investment turns into financial loss as the harvest (due to various reasons) failed to bring expected returns.

---

[4] In a qualitative study, the probing technique refers to stimulating an informant to produce more information.

4. Lack of female job—lack of jobs for women who are actively searching for a permanent job to support their family.
5. Credit burden—microcredit borrowed from NGOs where the defaulting rural borrowers are allegedly being harassed by NGO staff.
6. Loss in small business—unsuccessful outcome in small business like running a grocery store.
7. Drinkable water scarcity—prioritised as a critical driver by landless individuals who lack land for dwelling and farming.

10.4 Groundwater Irrigation to Combat Seasonal Drought or Just Moving the Problem?

Groundwater depletion and uneven recharge is a global concern, especially for countries based on agriculture. About 65% of all freshwater withdrawal worldwide is consumed by agricultural irrigation (Shamsudduha, Taylor, Ahmed, & Zahid, 2011). However, more than 25% of the world's total irrigation is supplied by groundwater, whereas 75% of these extractions take place in Asia (Shah, Burke, & Villholth, 2007). In Bangladesh, during 2004–2005, about 75% of the total estimated annual irrigation water came from groundwater sources by using various pumping techniques, including deep tube wells (Siebert et al., 2010). Recent studies (see e.g. Rodell, Velicogna, & Famiglietti, 2009 ; Shamsudduha, Chandler, Taylor, & Ahmed, 2009 ; Tiwari, Wahr, & Swenson, 2009) warn of a consistently high declining trend in the depth of groundwater (0.1–0.5 m/year) in some South Asian countries like India and Bangladesh. Indeed, they represent the world's second and fourth largest rice-producing nations, respectively, and both are highly dependent on groundwater-fed irrigation to cultivate high-yielding rice and other crops during the dry season (IRRI, 2010; Ranman & Mahbub, 2012; Scott & Sharma, 2009).

In Bangladesh, agricultural irrigation was dependent on traditional means without institutional intervention when famers relied on monsoon rainfall and surface water from nearby rivers and channels prior to the 1970s (UNDP, 1982). In contrast, presently 79% or more of agricultural land is irrigated by underground supply water during *boro* (a common crop variety) season (BADC, 2010). The formation of the East Pakistan Water and Power Development Authority took place in 1959 (Ali, 2011), now known as the Bangladesh Water Development Board. In order to standardise the ground and surface water irrigation programme, the East Pakistan Agricultural Development Corporation was formed in 1961, which later became known as Bangladesh Agricultural Development Corporation (BADC) (IBRD, 1970). The BADC initiated the installation of 3000 deep tube wells in their northwest irrigation project across the country, to facilitate irrigation for farming. However, the high Barind region was excluded from this installation programme due to its low potential for a groundwater aquifer (BMDA, 2006). With the formation of the Barind Integrated Area Development Project in 1985 under BADC, which later became the

Barind Multi-purpose Development Authority (BMDA) in 1992, large-scale groundwater irrigation expanded in the Barind region (BMDA, 2011).

In Bangladesh, the annual average rainfall is approximately 2500 mm/year; for the Barind area it is below 1000 mm, so is less than a half of the country's average (Ranman & Mahbub, 2012). The soil of the Barind region is made of Pleistocene clay, locally known as *Madhupur* clay in the subsurface, which is very thick (Ahmed, 2006). This subsurface Pleistocene clay resists vertical infiltration of rainwater (Ahmed & Hossain, 2006), particularly in the high Barind area, where the clay thickness is up to 50 m (Ahmed, 2006) and thus the vertical infiltration of rainwater becomes relatively very low. Overall, due to low rainfall and thick subsurface clay, the recharge rate of groundwater is very low in the Barind area.

At a depth of about 200–250 feet from the surface there is no underground aquifer. Instead, hard clay remains in most of the underground layers of the Barind soil. Hence, both the drinking and irrigation are dependent on the upper aquifer, i.e. from surface to 200/250 feet. The next aquifer exists 700–800 feet from the surface. Additionally, the Barind area does not have many rivers compared to other parts of the country that could be useful for surface water irrigation or groundwater recharge. Historically, this has caused crop failure and seasonal drought, as groundwater irrigation with an irrigation pump is not easy in the Barind area (BMDA, 2006).

The main reason behind the formation of the BMDA was to combat seasonal drought and crop failure by offering groundwater irrigation, mainly by drilling deep tube wells (BMDA, 2011). Under this initiative, groundwater extraction for irrigation expanded by drilling thousands of deep tube wells across the region. Expanded groundwater irrigation together with the introduction of high-yielding variety paddy seeds has magnified crop production—a so-called "green revolution". Due to the high volume of groundwater extraction from the upper aquifer using borewells, the depth of groundwater is rapidly increasing (Ranman & Mahbub, 2012). Moreover, in most of the Barind area, the groundwater is not being recharged during the shorter monsoon rainfalls (Ranman & Mahbub, 2012). Figure 10.1 shows the consistently increasing average depth of the groundwater between the years 2000 and 2014 in the Tanore Sub-district.

Due to the depth of the groundwater, the traditional hand-pumped tube wells were not useful for extracting groundwater for household consumption. Eventually the shortage of drinking water created conflict between the BMDA and the local community, according to the government's key informant in this study. In order to mitigate the conflict, the BMDA extended their piped water supply network in many locations to the community level. To date, a list of environmental and social awareness programmes have taken place against the huge groundwater depletion project. Nevertheless, recently BMDA started supplying a portion of irrigation water from the surface, especially from the northern rivers like the Ganges and Mahananda. After the direct order from the Ministry of Agriculture, the BMDA stopped drilling new borewells in 2013, as updated by the same key informant. Ironically, the local people have already adopted the technology of drilling deep tube wells for groundwater irrigation despite a ban from the government. The cost of such a private irrigation facility, illegally supplied by local groups, is much higher than the price charged

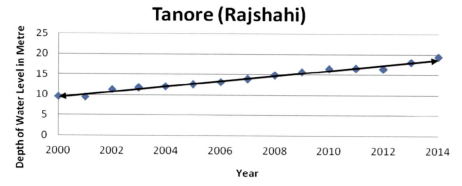

Fig. 10.1 Depth of groundwater level at Tanore sub-district: years 2000–2014 (Source: depth of groundwater captured at six different BMDA observatories)

by the BMDA system. Usually, farmers have to pay between 110 and 130 Bangladeshi taka (approximately US$1.50) to BMDA for one hour of groundwater irrigation system to the crop field. The cost of the "private" irrigation system is almost double, and very difficult for marginal (landless) farmers to bear, as informed by the participants of this study.

10.5 Results and Discussion

For this study, the types of migration were divided into two categories: short-term and long-term migration. In total, 8 short-term and 18 long-term migration incidents are documented.

Short-term migrants are individuals who have moved out from their geographical place of origin and are staying in other locations for a short period, i.e. less than a year. For this study, short-term migration also includes seasonal migration—people who migrate during seasonal unemployment and who usually return home when income opportunities are restored at their place of origins (e.g. agricultural labourers). Long-term migrants are those who have migrated elsewhere and are staying there for more than a year. For simplification, the definition of long-term migration is extended to include permanent migrants who have no plans to return to their place of origin. In both types of migration, the minimum distance that the migrants travelled was at least to another sub-district within the same district or any other districts within the country. Most of the migration happened within the internal geographical location of the country and mainly involved two categories: mobility within the same district and mobility towards Dhaka Division. (The administration of Bangladesh is divided into eight major regions called divisions, with Dhaka Division having the capital city as the centre.) One exception was where long-term mobility involved crossing the country border to India, as mentioned by the relatives of a migrant.

10.5.1 Description of the Critical Drivers for the Migration Threshold

Each of the study participants was asked to explain the critical driver for migration in order to understand the threshold points of outmigration in the disadvantaged Barind area. The most frequent critical driver (Table 10.1) is "seasonal lack of income", reported in 9 out of 26 interviews. Those who suffer from a seasonal lack of income are mainly marginal farmers, fishermen and agricultural labourers who migrate from geographical origins to elsewhere to earn wages during seasonal crises. These short-term seasonal migrants basically adopt secondary occupational skills at their destinations, taking jobs such as a rickshaw puller, construction worker or agricultural labourer (Table 10.1). Every year, most of these short-term migrants tend to return to their geographical origin when the local income opportunity is restored. For example, a marginal (i.e. sharecropping or subsistence) farmer who goes to a city for rickshaw pulling returns home during the time of harvesting and planting new agricultural crops. Other household members take care of the farm during his absence.

The second-highest identified critical driver to explain migration threshold is micro-"credit burden", which involves significant tension between poor money borrowers and the local NGO staff who are running rural microfinance projects. For various reasons, the rural poor are in need of urgent credit, to purchase agricultural inputs, run small business, attend emergency medical treatment, recover from damage after a natural calamity (e.g. storms), etc. Borrowing money from local NGOs is more accessible for rural people than approaching traditional or noninstitutional money lenders. In both lending sources, the credit procedure imposes higher interest rates on weekly, monthly or annual instalments. Local money lenders usually charge exorbitant interest rates, ranging from 120 to 200%, and take mortgage of valuables like gold, jewellery and land. Microfinance institutions in Bangladesh charge 40–120% interest which is also not helpful for the poor (Angelucci, Karlan, & Zinman, 2013). However, the microcredit schemes run by NGOs do not require collateral, since they claim to be based on social collateral. Social collateral refers to the social ties between the borrower group members. Borrowers are often linked to nonborrowers from their community, and there is the threat of social sanctions through loss of trust and embarrassment in case of failure to repay a loan (Aslam & Azmat, 2012).

The success of microfinance projects worldwide is measured by high credit recovery rates (Angelucci et al., 2013). Many NGOs in Bangladesh appoint a female applicant's husband as guarantor for the money while registering the woman as the principal borrower. Study participants alleged that NGO staff go door to door and harass individuals who have not regularly repaid the promised weekly or monthly instalments. In extreme cases, multiple staff arrive to intimidate and verbally and physically abuse the person/s who have not repaid monies; within the 26 cases, 4 respondents alleged such abuse by NGO staff. Nonetheless, studies like Chowdhury (2009) and Angelucci et al. (2013) have suggested that microfinance is not a cure-all

10 Seasonal Drought Thresholds and Internal Migration for Adaptation: Lessons... 179

Table 10.1 Summary of the studied migrants and the critical drivers to explain migration thresholds

Respondent ID	Male/female	Union as geographic origin	Critical driver to explain migration threshold	Type of migration	Occupation at geographic origin	Occupation at geographic destination	Geographic destinations
1	Male	Chanduria	Seasonal lack of income	Short term	Farmer	*Raj Mistry*	Within Dhaka Division
2	Male	Chanduria	Seasonal lack of income	Short term	*Raj Mistry*	*Raj Mistry*[a]	Within the same district
3	Male	Chanduria	Seasonal lack of income	Short term	*Raj Mistry*	*Raj Mistry*	Within Dhaka Division
4	Male	Chanduria	Seasonal lack of income	Short term	Farmer	*Raj Mistry*	Within the same district
5	Male	Chanduria	Seasonal lack of income	Short term	Agricultural labourer	Rickshaw puller[b]	Within Dhaka Division
6	Female	Chanduria	Lack of female job	Long term	Housewife	Garments worker	Within Dhaka Division
7	Male	Talondo	Credit burden	Long term	Farmer	Garments worker	Within Dhaka Division
8	Male	Talondo	Seasonal lack of income	Long term	Farmer	Garments worker	Within Dhaka Division
9	Male	Talondo	Credit burden	Long term	Agricultural labourer	Garments worker	Within Dhaka Division
10	Female	Talondo	Credit burden	Long term	Housewife	Factory labourer	Within Dhaka Division
11	Male	Talondo	Credit burden	Long term	Rickshaw van puller	Garments worker	Within Dhaka Division
12	Male	Talondo	Credit burden	Long term	Agricultural labourer	Factory labourer	Within Dhaka Division
13	Male	Chanduria	Loss in agriculture	Long term	Farmer	Construction labourer	Within the same district
14	Female	Chanduria	Lack of female job	Long term	Salesman	Rickshaw puller	Within Dhaka Division
15	Male	Chanduria	Lack of income	Long term	*Raj Mistry*	Factory Labourer	Within Dhaka Division

Table 10.1 (continued)

Respondent ID	Male/female	Union as geographic origin	Critical driver to explain migration threshold	Type of migration	Occupation at geographic origin	Occupation at geographic destination	Geographic destinations
16	Female	Chanduria	Lack of female job	Long term	Housewife	Service	Within the same district
17	Male	Chanduria	Credit burden	Long term	Farmer	Garments worker	Within Dhaka Division
18	Female	Chanduria	Lack of female job	Long term	No occupation	Garments Worker	Within Dhaka Division
19	Female	Chanduria	Lack of female job	Long term	Student	Factory labourer	Within Dhaka Division
20	Male	Talondo	Loss in small business	Long term	Cook	Cook	Within the same district
21	Female	Tanore	Lack of female job	Long term	No occupation	Garments worker	Within Dhaka Division
22	Male	Tanore	Seasonal lack of income	Short term	Fisherman	Construction labourer	Within the same district
23	Male	Tanore	Seasonal lack of income	Short term	Fisherman	Construction labourer	Within the same district
24	Male	Tanore	Seasonal Lack of Income	Short term	Fisherman	Construction labourer	Within the same district
25	Male	Tanore	Credit burden	Long term	Small business	Unknown	Cross boarder - to India
26	Male	Badhaile	Drinkable water scarcity	Long term	Farmer	Factory labourer	To a nearby sub-district

[a]Raj Mistry or Raj Mistri is a term used in the Indian subcontinent that refers to those who are master craftsmen, expert masons and also bricklayers
[b]A rickshaw is a light passenger tricycle powered by human pedalling, commonly seen in Bangladesh

for poverty reduction. Rather, an impact analysis of microfinance concluded that no transformative impacts but more positive than negative impacts. Borrowers who already have some assets are more likely to have a successful repayment rate than relatively vulnerable groups like widows and landless people (Chowdhury, 2009; Jain & Moore, 2003).

The marginal rural poor in Barind have continuous financial shortfall and are often forced to use the borrowed money to solve other immediate expenses instead of investing it in income-generating activities/ventures, such as starting a small business, buying fertiliser or paying for groundwater irrigation of their agricultural land. When asked how the microcredit was used, four study participants reported that while the official reason for borrowing money was to invest in small business and agriculture, the money has been partly spent to repair their houses; nine respondents commented that the previous year's hailstorms damaged their house and monies were need for these repairs; three respondents reported using all the allocated microcredit received to meet urgent medical expenses including the cost of a childbirth; and one respondent experienced a recent loss after investing the borrowed money on a seasonal business. Additionally, some of the respondents claimed they were already in debt to other NGOs, had been intimidated by other credit recovery staff and had repaid other outstanding loan amounts from the new monies borrowed. This difficult cycle of debt leads to more poverty and decisions to out-migrate or pursue alternative livelihoods.

Bangladesh microcredit institutions offer loans to disadvantaged women and other poor individuals with the objective to meet global development agendas (World Bank, 2014). According to government records, until February 2014, the country had 2457 registered local and international NGOs working in rural and urban areas (NGOAF, 2014). Due to a lack of screening and monitoring systems, the same individual can be considered for several microcredit schemes (run by different NGOs) at the same time. The study interviews disclosed that 9 out of 26 respondents concealed their credit situation where monies were owed or previous loans had defaulted with NGOs, while at the same time applying for new microcredit. Indeed, the interviews demonstrated that about one-third of respondents had received loans from more than one NGO, exponentially increasing their credit burden and inability to repay instalments.

10.5.1.1 Case Study Interview: Male Aged 47 at Talondo Union

Mr X, a male aged 47, is a landless farmer (locally known as "Adhi Borga") who leases land for agriculture. The entire production cost is borne by Mr X. As per his contract, after harvesting, half of the crop/income from the paddy should be given to the owner of the land. Due to some illness in his family, some of the NGO microcredit money was spent to bear a medical expense and the rest was spent to adjust existing loan repayments with another NGO credit loan. The following statement of Key Informant 7 (i.e. Mr X's wife) further illustrates the link between the vicious cycle of poverty and institutional credit burdens, and is an important critical driver discovered in this study.

> That was the dry season and the cost of irrigation was high but a portion of the NGO's money was spent for my medical treatment because I was ill and hospitalised. Consequently, we failed to pay hourly for irrigation and our land was half irrigated and resulted in a bad harvest. We could not pay the debt taken from the NGO. Their staff members abused us verbally by frequently visiting our home. Due to this "shame", my husband ran away to Dhaka for a job and is now working in a garments factory.

In summary, other critical drivers (Table 10.1) include "lack of female employment/job", "loss in small business", "drinkable water scarcity" and "lack of income" (year-round). "Lack of female job" is experienced by rural women who have completed primary education at schools (this study observed that educational attainment varies between grades 8 and 12). In most cases, these young women became involved in the job market in response to long-term illness or death of male-earning members of their family. This group of female migrants is not willing to join seasonal labour jobs at the local level and is attracted to work in garments and other factories in Dhaka Division, including Gazipur Shavar and Dhaka districts. The critical drivers around lack of income and drinkable water scarcity appeared only once in Table 10.1 and thus are considered atypical in this case study.

10.5.2 Presence of Environmental Drivers in the Migration Threshold

Although the study participants initially claimed vulnerability in terms of their financial situation, the significant influence of seasonal drought and occasional monsoon floods were the main environmental drivers within this complex environment–poverty nexus.

For seven interviewees, environmental factors were linked to the critical driver that led to their final decision to migrate (Table 10.1). Mr Y, a 42-year-old landless farmer (respondent number 13 in Table 10.1), initiated sharecropping in a low-lying area prone to monsoon flooding. As the Barind region contains mostly elevated ground compared to other parts of the country, the low-lying areas in Tanore Sub-district are mainly situated beside riverbanks. Unfortunately, Mr Y's crop field was inundated by flood during the monsoon. Although the unavoidable financial "loss in agriculture" is accentuated as his critical driver to make the final decision to migrate (Table 10.1), the lowland flood damage has been an environmental driver inducing such financial consequences, as understood from Mr Y's interview. Similarly, two other participants (respondents 7 and 17 in Table 10.1) who were basically landless farmers have linked seasonal drought (which is an environmental driver) as inducing a bad harvest in sharecropping. The landless farmers' failure to repay the microcredit (intended originally to bear the cost of farming) instalments ultimately exposed them to alleged harassment by the money-lending institutions. With no other option left, both males migrated quietly away from their place of origin. In due course, "credit burden" has seemingly become the critical driver for migration.

Nonetheless, this study also interviewed fishermen (respondents 22–24 in Table 10.1) who usually migrate for 4–6 months of the year due to lack of fishing opportunities in their locality. First, in the dry season, fishing is not feasible as the rivers and channels dry out. Second, the interviewed fishermen have claimed that changing weather conditions, like a longer seasonal drought in recent years, are linked to the shortfall of water flow and unproductive fishing in local river channels. The dry season persists roughly from November to April each year, so fishermen suffer from lack of income locally and migrate seasonally to other places for alternative livelihoods.

Nevertheless, one atypical case was found where a family's migration decision was driven by drinkable water scarcity at the geographical origin (respondent 26 in Table 10.1). The family lived in Badhaile Union, one of the highest elevated areas of the Barind region. The key informant interviews with local government representatives found that the BMDA technicians had repetitively failed to reach any underground aquifer in Badhaile Union. Despite low population density, agriculture in Badhair Union is solely depended on monsoon rain producing a single crop each year. Every day, household members have to carry drinkable water for up to 3 km. Not only do they pay money to purchase piped reticulated water, it becomes difficult for women to carry water during the rainy season. Here the critical driver for migration was "drinkable water scarcity" at the geographical origin. However, the migration decision was validated by an opportunity to access government *Khas* land at the new geographical destination in a nearby sub-district.

10.6 Conclusion and Recommendations

The study examined how the disadvantaged rural population living in the northern drought-prone areas of Bangladesh make decisions to migrate to adapt to environmental and nonenvironmental stressors. While the current research's emphasis is to establish the linkage between slow-onset environmental hazards and human migration with systematic evidence (also see in Gray & Mueller, 2012), the issue will impose greater risks for many developing countries that are being exposed to climate change impacts. The case study approach allowed in-depth, multifaceted explorations of complex issues influencing the migration decision in their real-life settings.

The concept of a threshold point for migration has been accepted in this study for a better understanding of an individual's breaking point before outmigration. In order to describe the migration threshold points, the critical drivers (Table 10.1) or the respondents' final reasons for migration have been examined. In addition, other drivers of vulnerability involving socioeconomic and environmental components have been considered. Besides in-depth interviews with the migrants' households, key informant interviews and focus group discussions were conducted with the participants residing in this northern seasonal drought-prone sub-district of Bangladesh.

Qualitative data were obtained using semi-structured questionnaires complemented by a probing technique used in communities with low literacy skills; the conversations were in the local language and later transcribed to English (see Sect. 10.3).

Results support the fact that the disadvantaged rural poor are exposed to a range of emerging drivers of vulnerability involving both environmental and nonenvironmental components. The key environmental drivers include seasonal drought, lack of surface water for irrigation, lack of drinkable water/water scarcity and impacts of storms, whereas the nonenvironmental components comprise lack of income causing poverty, increasing cost burden of medical treatment, death of the earning family member, higher burdens of microcredit, etc. The disadvantaged households always have a financial shortfall where both their coping strategy and adaptation to environmental and nonenvironmental shocks are compromised. This is probably the main reason why most of the critical drivers of migration are reported in financial terms (e.g. seasonal lack of income). For this community, one way to solve a financial shortfall is to borrow microcredit from the local NGOs. Since most of the respondents have taken high levels of repayable NGO credit and at a high installment rate, this situation creates increased financial strain on already economically and environmentally vulnerable individuals and groups in the studied area of Bangladesh.

One clear reported impact was the NGOs' continual harassment of individuals and households when attempting to recoup unpaid instalments. An unexpected but major finding of this research is the stress from and large number of participants identifying microcredit burden as the critical driver influencing their outmigration decision.

This study also discovered that in the context of slow-onset hazards like seasonal drought, unpacking environmental drivers separately from those which are seemingly nonenvironmental ones (e.g. economic and social drivers) is not relevant. Generally, critical drivers are complex and are compounded by the impacts of other drivers; the consolidated effect of various vulnerabilities needs unpacking in a single livelihood equation. Hence, in the context of slow-onset hazards like drought, the threshold of a migration point may not be as simple as understanding one critical driver.

A limitation of this study is that the case study focused specifically on a subdistrict ranked as the most vulnerable to seasonal drought. Different locations with sociogeographical variability may not have similar compounding factors and so perhaps may have different findings. For example, remote locations like the Sand Islands beside the flooding rivers catchments (locally known as *char*) in northern Bangladesh may not be widely exposed to government and nongovernment intervention projects like the studied location. Further studies should explore the dynamics of critical drivers for rural–urban migration at those locations.

Apart from its conceptual contribution, this study offers some policy implications on behalf of the marginal groups living in the Barind Tract area. With the formal expansion of projects by government and other unauthorised deep tube-well owners, access to water both for drinking and irrigation has become dependent on individuals' ability to pay. This competition for water will worsen when the underground water level further decreases (Fig. 10.1). Landless farmers, especially those who are surviving on sharecropping, are already in serious financial distress. This is an issue where immediate policy interventions are required. Since the BMDA is the leading government authority to ensure agricultural irrigation facilities to the high Barind area, the key policy decision-makers should consider two questions: whether charging money for agricultural irrigation is an essential policy approach and the existing high rate per hour irrigation needs review, and if water should be distributed in a more equitable way so that every farmer is ensured access to irrigation? Ongoing internal outmigration is a consequence of these urgent policy decisions.

Further to this question, groundwater irrigation in any case is not a feasible option for the longer term in this Barind area considering the increasing depth of groundwater. Emphasis should then be given to using surface water for irrigation, with discussion on transboundary water sharing with the neighbouring country India. Further research is required; however, it was reported during interviews that the immediate local solution includes strengthening alternative technologies like artificial recharge of groundwater, rain water harvesting, and possibly making a barrage to supply water to ensure irrigation in the dry Barind area.

Nonetheless, microcredit has commonly become the last resort for the rural poor to combat vulnerability. The rural poor often take credit from multiple sources to adjust previous borrowings. However, in the long run, such a practice may worsen the situation with a large burden of repayment instalments and unpleasant confrontations with the NGOs and their representatives. In extreme cases, individuals are displaced for the long term, switching their occupation from farming to work mostly as labourers (in factories and construction sites). Migrated individuals conceal their current address and it is increasingly harder for NGOs to recoup outstanding loans. The intention of this argument is not to ban microcredit options, but a serious review of the microcredit scheme at the grassroots levels is essential. The key solution is for NGO credit policymakers, local credit managers and credit field staff to consider other less conflicting ways to recover unpaid credit (Datar, Epstein, & Yuthas, 2008). Indeed, the success of microcredit programmes should be redefined on the basis of their qualitative contribution to the livelihood of credit beneficiaries, not in terms of high credit recovery rates.

References

Ackah, C., & Medvedev, D. (2012). Internal migration in Ghana: Determinants and welfare impacts. *International Journal of Social Economics, 39*(10), 764–784. https://doi.org/10.1108/03068291211253386.

Adams Jr. R. H., Cuecuecha, A., & Page, J. (2008). The impact of remittances on poverty and inequality in Ghana. *The Would Bank Policy Research Working Paper 4732*. Retrieved from documents. worldbank.org/curated/en/393501468030835717/pdf/

Afsar, R. (2000). *Rural urban migration in Bangladesh: Causes, consequences and challenges.* Dhaka: University Press Limited.

Afsar, R. (2003). *Internal migration and the development nexus: The case of Bangladesh*. Paper presented at the Regional Conference on Migration, Development and Pro-Poor Policy Choices in Asia, 22–24 June, Dhaka; Refugee and Migratory Movements Research Unit, Dhaka, and UK Department for International Development, London.

Ahmed, K. M. (2006). *Barind Tract*. Retrieved July, 2013 from http://www.banglapedia.org/httpdocs/HT/B_0309.HTM

Ahmed, K. M., & Hossain, M. A. (2006). *Groundwater*. Retrieved June 13, 2016 from http://www.banglapedia.org/httpdocs/HT/G_0209.HTM

Ahsan, R., Karuppannan, S., & Kellett, J. (2011). Climate migration and urban planning system: A study of Bangladesh. *Environmental Justice, 4*(3), 163–170. https://doi.org/10.1089/env.2011.0005.

Ali, A. M. (2011). *Fundamentals of irrigation and on-farm water management* (Vol. 1, p. 18). New York: Springer.

Angelucci, M., Karlan, D., & Zinman, J. (2013). *Microcredit impacts: Evidence from a randomized microcredit program placement experiment by Compartamos Banco*. Retrieved August 1, 2016 form http://www.dartmouth.edu/~jzinman/Papers/CompartamosImpact_Dec16_2013.pdf

Aslam, A., & Azmat, N. (2012). *A study of collateral options for microfinance loans in Pakistan*. Pakistan Microfinance Network. Retrieved February 18, 2017 from https://www.microfinancegateway.org/

BADC (Bangladesh Agricultural Development Corporation). (2010). *Minor Irrigation Survey Report 2009–10* (p. 3). Dhaka: BADC.

Bardsley, D. K., & Hugo, G. (2010). Migration and climate change: Examining thresholds of change to guide effective adaptation decision-making. *Population and Environment, 32*, 238–262. https://doi.org/10.1007/s11111-010-0126-9.

BBS (Bangladesh Bureau of Statistics). (2011). *Population and household census 2011*. Retrieved from http://www.bbs.gov.bd/

Beegle, K., de Weerdt, J., & Dercon, S. (2011). Migration and economic mobility in Tanzania: Evidence from a tracking survey. *The Review of Economics and Statistics, 93*(3), 1010–1033. https://doi.org/10.1162/REST_a_00105.

Biermann, F., & Boas, I. (2010). Preparing for a warmer world: Towards a global governance system to protect climate refugees. *Global Environmental Politics, 10*(1), 60–88. https://doi.org/10.1162/glep.2010.10.1.60.

BMDA (Barind Multipurpose Development Authority). (2006). *Borandro authority past-present* (p. 35). Bangladesh: Rajshahi.

BMDA (Barind Multipurpose Development Authority). (2011). *Barind multipurpose development authority: Overview*. Retrieved from http://www.bmda.gov.bd/

Boano, C., Zetter, R., & Morris, T. (2007). *Environmentally displaced people: Understanding the linkages between environmental change, livelihood and forced migration*. *Oxford*. Oxford, UK: Refugee Studies Centre.

Brooks, N., Adger, W. N., & Kelly, P. M. (2005). The determinants of vulnerability and adaptive capacity at the national level and the implications for adaptation. *Global Environmental Change, 15*(2), 151–163.

CDMP (Comprehensive Disaster Management Program). (2013). *Drought vulnerability ranking in Bangladesh*. Ministry of Disaster and Relief. Retrieved July, 2015 from www.cdmp.org.bd

Chowdhury, A. (2009). Microfinance as poverty reduction tool: A critical assessment. *Economic and Social Affairs. DESA Working Paper No. 89*. Retrieved August 15, from http://www.un.org/esa/desa/papers

CIA (Central Intelligence Agency) (2011). *The world fact book*. Retrieved June 29, 2011 from the original. Retrieved August 8, 2011.

Datar, S. M., Epstein, M. J. & Yuthas, K. (2008). In *Microfinance, clients must come first. Social innovations* (Winter 2008).

de Brauw, A., Mueller, M., & Woldehanna, T. (2013). *Does internal migration improve overall well-being in Ethiopia?* (Ethiopia Strategy Support Program (ESSP) Working Paper 55). Retrieved from http://www.ifpri.org/publication/does-internal-migration-improve-overall-well-being-ethiopia

Dickin, S. K., Schuster-Wallace, C. J., & Elliott, S. J. (2013). Developing a vulnerability mapping applying the water-associated disease in Malaysia. *PLoS One, 8*(5), e63584. https://doi.org/10.1371/journal.pone.0063584.

Fekete, A. (2009). Validation of a social vulnerability index in context to riverfloods in Germany. *Natural Hazards and Earth System Sciences, 9*, 393–403.

Findley, S. (1994). Does drought increase migration? A study of migration from rural Mali during the 1983–1985 drought. *The International Migration Review, 28*, 539–553.

Gemenne, F. (2013). Migration doesn't have to be a failure to adapt. In J. Palutik, S. L. Boulter, A. J. Ash, M. S. Smith, M. Parry, M. Waschka, & D. Guitart (Eds.), *Climate adaptation futures* (pp. 235–241). Oxford: Wiley. https://doi.org/10.1002/9781118529577.

Gray, C., & Mueller, V. (2012). Natural disasters and population mobility in Bangladesh. *Proceedings of the National Academy of Sciences of the United States of America, 109*(16), 6000–6005. https://doi.org/10.1073/pnas.1115944109.

Hagen-Zanker, H. (2008). *Why do people migrate? A review of the theoretical literature.* (MPRA Paper No. 28197). Retrieved from http://mpra.ub.uni-muenchen.de/28197/

IBRD (International Bank for Reconstruction and Development). (1970). IBRD-IDA tube well project, EPADC, East Pakistan, 1970, Vol. PA-49a. *Annex,11*, 1.

IOM. (2010a). *The World migration report 2010: The future of migration: Building capacities for change, Geneva.* Retrieved January 25, 2017 from http://publications.iom.int/bookstore/free/

IOM. (2010b). *Climate change, environment and migration: Frequently asked questions. Climate change, environment and migration alliance, December 2010.* Retrieved January 25, 2017 from http://www.iom.int/jahia/webdav/shared/shared/mainsite/activities/

IOM (International Organization for Migration). (2009). *Migration, environment and climate change: Assessing the evidence.* Retrieved August 27, 2016 from http://publications.iom.int/system/files

IRRI (International Rice Research Institute). (2010). *World rice statistics (WRS), Manila, Philippines.* http://beta.irri.org/index.php/Social-Sciences-Division/SSD-Database/. Cited April 2010

Jain, P., & Moore, M. (2003). *What makes Microcredit Programme Effective? Fashionable fallacies and workable realities* (IDS Working Paper 177). Brighton : Institute of Development Studies, University of Sussex.

Kabir, M. E. (2013). Shrimp cultivation and feminist environmentalism. *Journal of Public Administration. Dhaka, 21*(1), 63–76.

Kartiki, K. (2011). Climate change and migration: A case study from rural Bangladesh. *Gender and Development, 19*(1), 23–38. https://doi.org/10.1080/13552074.2011.554017.

Khandker, S. R. (2012). Seasonality of income and poverty in Bangladesh. *Journal of Development Economics, 97*(2), 244–256. https://doi.org/10.1016/j.jdeveco.2011.05.001.

Kniveton, D., Schmidt-Verkerk, K., Smith, C., & Black, R. (2008). *Climate change and migration: Improving methodologies to estimate flows, Migration Research Series 33.* Geneva: International Organization for Migration.

Koubi, V., Spilker, G., Schaffer, L., & Bernauer, T. (2016). Environmental stressors and migration: Evidence from Vietnam. *World Development, 79*, 197–210. https://doi.org/10.1016/j.worlddev.2015.11.016.

Mallick, B., & Vogt, J. (2014). Population displacement after cyclone and its consequences: Empirical evidence from coastal Bangladesh. *Natural Hazards, 73*, 191–212. https://doi.org/10.1007/s11069-013-0803-y.

Michalos A. C., Smale, B., Labonte, R., Muharjarine, N., Scott, K. Guhn, M. ... Hyman, I. (2011) *The Canadian Index of Wellbeing* (Technical Report 1.0). Waterloo, ON.

Murray, V., McBean, G., Bhatt, M., Borsch, S., Cheong, T. S., Erian, S., et al. (2012). Case studies in managing the risks of extreme events and disasters to advance climate change adaptation.

In C. B. Field, V. Barros, T. F. Stocker, D. Qin, D. J. Dokken, K. L. Ebi, et al. (Eds.), *A special report of working groups I and II of the intergovernmental panel on climate change* (pp. 487–542). Cambridge: Cambridge University Press.

NGOAB (NGO Affairs Bureau of Bangladesh). (2014). *List of NGOs in Bangladesh*. Retrieved August 14, from ngoab.portal.gov.bd

Qin, H. (2010). Rural-to-urban labour migration, household livelihoods, and the rural environment in Chongqing municipality, Southwest China. *Human Ecology, 38*(5), 675–690. https://doi.org/10.1007/s10745-010-9353-z.

Ranman, M. M., & Mahbub, A. Q. M. (2012). Groundwater depletion with expansion of irrigation in Barind Tract: A case study of TanoreUpazila. *Journal of Water Resource and Protection, 4*, 567–575.

Reuveny, R. (2007). Climate change-induced migration and violent conflict. *Political Geography, 26*(6), 656–673.

Rodell, M., Velicogna, I., & Famiglietti, J. S. (2009).Satellite-based estimates of groundwater depletion in India. *Nature*. doi:https://doi.org/10.1038/nature08238

Scott, C. A., & Sharma, B. (2009). Energy supply and the expansion of groundwater irrigation in the Indus-Ganges Basin. *International Journal of River Basin Management, 7*, 1–6.

Shah, T., Burke, J., & Villholth, K. (2007). Groundwater: A global -assessment of scale and significance. In D. Molden (Ed.), *Water for food water for life: A comprehensive assessment of water management in agriculture* (pp. 395–423). London: Earthscan.

Shamsudduha, M., Chandler, R. E., Taylor, R. G., & Ahmed, K. M. (2009). Recent trends in groundwater levels in a highly seasonal hydrological system: The Ganges-Brahmaputra-Meghna Delta. *Hydrology and Earth System Sciences Discussions, 13*, 2373–2385.

Shamsudduha, M., Taylor, R. G., Ahmed, K. M., & Zahid, A. (2011). The impact of intensive groundwater abstraction on recharge to a shallow regional aquifer system: Evidence from Bangladesh. *Hydrogeology Journal, 19*, 901–916.

Siebert, S., Burke, J., Faures, J. M., Frenken, K., Hoogeveen, J., Döll, P., & Portmann, F. T. (2010). Groundwater use for irrigation: A global inventory. *Hydrology and Earth System Sciences Discussions, 7*, 3977–4021.

Speare, A., Jr. (1974). Residential satisfaction as an intervening variable in residential mobility. *Demography, 11*, 173–188.

SRDI (Soil Resource Development Institute). (2000). *Upazila land and soil resource utilization guide: Tanore, Rajshahi* (p. 10). Bangladesh: SRDI.

Tate, E. (2012). Social vulnerability indices: A comparative assessment using uncertainty and sensitivity analysis. *Natural Hazards, 63*, 325–347.

The World Bank. (2014). *Household survey to conduct micro-credit impact studies: Bangladesh*. In R. Shahidur, S. R. Khandker, and S. A. Samad, (Ed.). WB Policy Research Working Paper 6821. Retrieved April 19, 2017 from http://documents.worldbank.org/curated/en/456521468209682097/Dynamic-effects-of-microcredit-in-Bangladesh

Tiwari, V. M., Wahr, J., & Swenson, S. (2009). Dwindling groundwater resources in northern India, from satellite gravity observations. *Geophysical Research Letters, 36*, L18401. https://doi.org/10.1029/2009GL039401.

UN (United Nations). (2015). *World statistics pocketbook, Bangladesh*. Retrieved September 28, 2015, from http://data.un.org/CountryProfile.aspx?crName=Bangladesh

UNDESA—DSD. (2011). *Environmentally induced migration and sustainable development: Background paper*. Retrieved form http://www.webmeets.com/files/papers/eaere/2011/1107/Environmentally%20Induced%20Migration%20and%20Sustainable%20Development_Milan_Areikat_Afifi.pdf

UNDP (United Nation Development Programme). (1982). *Ground-water survey: The hydrogeological conditions of Bangladesh* (Technical Report, UNDP, New York, DP/UN/BGD-74-009/1).

Vrij, A., Van der Steen, J., & Koppelaar, L. (1994). Aggression of police officers as a function of temperature: An experiment with the Fire Arms Training System. *Journal of Community & Applied Social Psychology, 4*(5), 365–370.

Wolpert, J. (1966). Migration as an adjustment to environmental stress. *The Journal of Social Issues, 22*(4), 92–102.

Wouterse, F. (2010). Remittances, poverty, inequality and welfare: Evidence from the central plateau of Burkina Faso. *The Journal of Development Studies, 46*(4), 771–789. https://doi.org/10.1080/00220380903019461.

Chapter 11
Incentives and Disincentives for Reducing Emissions under REDD+ in Indonesia

Fitri Nurfatriani, Mimi Salminah, Tim Cadman, and Tapan Sarker

Abstract This chapter explores the fiscal incentives and disincentives that contribute either positively or negatively to reducing emissions from deforestation and forest degradation (REDD+) in Indonesia. Indonesia is an important participant in the UN Framework Convention on Climate Change programme on REDD+. The programme is funded through financial contributions from developed to developing countries, which can eventually be part of a country's nationally determined contribution to reducing emissions, either domestically, or via international emissions trading. Our study finds that there are a number of formal charges, fees and taxes that apply on forest-related activities in Indonesia, which are stipulated within regulations promulgated by various government departments. A range of informal subnational charges also apply to forest-related activities, which has often provided a monetary incentive for local government, especially forest-rich districts, to exploit their timber resources. However, this has been proven as a disincentive for REDD+ implementation in Indonesia. We also find that there is a need for improved financial governance in future fiscal policy reform, which should include the removal of perverse incentives for forest conversion, the equitable and accountable distribution of financial incentives, the prevention of corruption and fraud, and the strengthening of economic benefits for smallholders. We recommend that in implementing the REDD+, the Government of Indonesia should consider providing incentives for the nonexploitation of forests by businesses engaged in the provision of environmental services as well as carbon transactions. This could

---

The research is funded by ACIAR, Project: FST/2012/040 (Enhancing smallholder benefits from Reduced Emissions from Deforestation and Forest Degradation in Indonesia)

F. Nurfatriani • M. Salminah
Research & Development Centre for Social Economics, Policy, & Climate Change, Ministry of Environment & Forestry, Government of Indonesia, Gunung Batu No 5, Bogor, West Java 16118, Indonesia

T. Cadman (✉)
Institute for Ethics, Governance and Law, Griffith University, Nathan, QLD, Australia
e-mail: t.cadman@griffith.edu.au

T. Sarker
Griffith Centre for Sustainable Enterprise and Department of International Business and Asian Studies, Griffith University, Nathan, Queensland, Australia

© Springer International Publishing AG 2018
M. Hossain et al. (eds.), *Pathways to a Sustainable Economy*,
DOI 10.1007/978-3-319-67702-6_11

take the form of private investments, private–public partnerships or civil society engagement in forestry and land use change, and may include incentives such as payment for ecosystem services and for forest ecosystem restoration.

Keywords Fiscal incentives and disincentives • REDD+ • Climate change • Indonesia

11.1 Background: REDD+, Institutional Arrangements and the Indonesian Context

The United Nations Framework Convention on Climate Change (UNFCCC) mechanism "reducing emissions from deforestation and forest degradation and the role of conservation, sustainable management of forests and enhancement of forest carbon stocks in developing countries", referred to as REDD+, is an institutional complex made up of various intergovernmental and national elements, with varying degrees of collaboration (Lubell, 2015, pp. 41–47). The mechanism is still evolving, reflecting the policy machinations of the UNFCCC out of which it emerged, and is made up of a mixture of intergovernmental and national practices (Palmujoki & Virtanen, 2016, pp. 59–78). Previous experiences with REDD+ have been largely based on "nested" (i.e. subnational) demonstration or pilot projects, with involvement of non-government organisations and other non-state actors. The trend now appears to be one of increasing centralisation and government control, with the danger that any financial benefits arising from emissions reduction payments will not flow to local actors, thereby reducing compliance. These issues, combined with inconsistent institutional approaches and a high level of policy uncertainty in the wake of the Paris Agreement, are contributing to a lack of clarity around what arrangements will ultimately be used to implement REDD+ on the ground (Vijge, Brockhaus, Di Gregorio, & Muharrom, 2016). At the intergovernmental policy level, there has been some recognition of the challenges confronted by national REDD+ initiatives. The Cancún Agreements of COP16 in 2010 refer specifically to the need for "guidance and safeguards" including "transparent and effective national forest governance" and the "full and effective participation of relevant stakeholders, in particular indigenous peoples and local communities", but they do not stipulate how these should be implemented, referring only to "national legislation and sovereignty" (UNFCCC, 2011) (p. 26). Under Cancún, tropical forest countries involved in the various stages of REDD+ (readiness, pilot studies and implementation) are expected to ensure national activities do not impact negatively on the environment or people, while also contributing positively to governance and generating environmental and social benefits (ICIMOD, 2016). In the wake of the Cancún Agreement, the REDD+ policy community responded by adopting a range of measures, notably around benefit-sharing arrangements. The World Bank's Forest Carbon Partnership Facility has played a significant role, requiring allocations from its Carbon Fund to occur in the context of a national benefit-sharing plan; however, exact arrangements are not specified (Chapman, Wilder, & Millar, 2014). In the case of stakeholder management, the World Bank requires increased and ongoing levels of engagement as and provides new standards for international financial institutions (Bank, 2015).

Private sector experiences in REDD+ engagement across Asia also reveal that major issues to implementation include unclear land tenure, regulatory uncertainty, bureaucratic processes and high transaction costs that impact on participation (Chokkalingam & Vanniarachchy, 2011). This has led to less involvement in the REDD+ process than other sectors, despite the critical role of business and industry in making sustainable land use decisions for REDD+ (Somorin, Visseren-Hamakers, Arts, Sonwa, & Tiani, 2014). More efforts should be directed in better understanding the role that finance can play in contributing to REDD+ through fiscal incentives, as recommended by Henderson, Coello, Fischer, Mulder, and Christophersen (2013). This includes engaging with and involving the private sector in a constructive discussion on REDD+, to understand what the enabling conditions are that would attract private sector involvement at scale.

This chapter explores the domestic-level fiscal instruments affecting the implementation of REDD+ activities in Indonesia. In Indonesia, REDD+ has been recognised as a potentially significant source of revenue, while at the same time providing an important incentive to contribute to reductions in global deforestation. The research explored the current performance of fiscal incentives, and primary industry and business stakeholder insights on whether the private sector can be encouraged to engage in REDD+ through fiscal incentives. This is important in order to determine practical and context-based approaches that will contribute to the overall aim of supporting the development of institutional arrangements and fiscal mechanisms for effective implementation of REDD+ within and among various levels of government, and ensuring that business and industry benefits from REDD+ in Indonesia.

11.2 REDD+ and Fiscal Instruments in Indonesia

11.2.1 History of REDD+

Intergovernmental efforts to reduce emissions from forest activities are closely tied to Indonesia and were formally established in 2007 as an output of the 13th Conference of the Parties (COP) to the UNFCCC in Bali under the Bali Action Plan. The concept first emerged at COP11 in Montreal in 2005, where Papua New Guinea and Costa Rica questioned the benefits of climate change mitigation actions for forest-rich developing countries under the Kyoto Protocol (with its emphasis on developed countries' obligations to mitigate emissions, albeit through developing countries and domestic action). Forests had previously been regarded as potential avenues for climate change mitigation by reducing emissions from deforestation (RED) arising from the conversion of forests into other land uses, with the further economic advantage that reductions in emissions would be less costly than in other sectors (Pan et al., 2011). Preliminary discussions were around creating a global and national incentive mechanism to encourage developing countries to avoid deforestation as well as forest degradation (REDD), and funding was provided by donor countries within the UNFCCC (Angelsen & McNeill, 2012). Subsequent negotiations led to the incorporation of conservation, the sustainable management of forests and carbon stock enhancement through afforestation and reforestation as co-benefits, resulting in the inclusion of the "plus" in REDD+ at COP15 in Copenhagen in 2009 and included at COP16 in 2010 in Cancún (Nurtjahjawilasa, Ysman, Septiani, & Lasmini, 2013).

The programme has three distinct phases: preparation (with the first projects commencing in 2007–2008), focused research, technology and policy development related to land use-based climate change mitigation; readiness (occurring over the period 2009–2012), which includes preparation in arranging REDD methodology and policies; and implementation (commencing in 2013). Preparation and readiness are important as they are related to the methodologies and incentive mechanisms agreed to between developing (recipient) countries and developed (donor) countries as the international climate negotiations move forward (Ministry of Forestry, 2010).

However, the incentive mechanisms for developing countries to adopt REDD+ remain unclear. The "results-based payment" approach refers to ensuring that emissions reductions have occurred against projected baseline emissions—the so-called requirement for "additionality" of mitigation activities, which would otherwise not have occurred (Kanninen et al., 2007; Righelato & Spracklen, 2007). Article 5 of the Paris Agreement reaffirms results-based payments for emissions reduction, but the mechanism is likely to face big challenges since countries now have to meet their own nationally determined contributions (NDCs) to reducing emissions and are likely to use emissions reductions from forests for their domestic reduction targets. Furthermore, Article 6 of the Agreement proposes a market mechanism for certified emissions reduction activities that allow carbon "offsets" between developing and developed countries in order to meet each country's NDC commitments under the UNFCCC. The nature of the incentive mechanisms to apply to REDD+ therefore continues to be under discussion in international climate negotiations. REDD+ activities have consequently stagnated, mostly due to unclear national regulations, the uncertainty of carbon markets and the lack of project funds at the present time. Also hanging over REDD+ is the question of whether it will continue to function with low transaction costs and high opportunity gains from carbon payments, as was initially assumed (Venter & Koh, 2012). There were high hopes that REDD+ would contribute to international emissions trading and make a significant contribution to reducing emissions (Eliasch, 2008; Stern, 2007). Other experts have expressed the view that carbon trading will merely shift emissions around the world and will become increasingly less effective in keeping global temperature below 2 °C, let alone the optimum target of 1.5 °C in the Paris Agreement (McAfee, 2016).

11.2.2 REDD+ in Indonesia

Deforestation and forest degradation in Indonesian tropical rainforests arises from a range of land uses including alienation of certain areas within designated forest areas; using forest areas for nonforestry purposes (e.g, palm oil plantations); maximising production forests for logging; developing community forest schemes in forest areas; plantation and agroforestry activities; exploitation of vacant land; and reconversion from other uses to native forest (Sahide & Giessen, 2015). With more than 130 million hectares of forest, Indonesia contains the world's third largest area of tropical rainforest and is critical to the success of REDD+ at the global level (REDD+ Task Force, 2012b). Through its National Action Plan on Greenhouse Gases Emission Reduction and the Mitigation Fiscal Framework, the country has set out its plan to implement its commitment to reduce emissions by 26% by 2020 compared to business as usual (Ministry of Finance, 2012).

In order to make REDD+ implementation compliant with the international consensus, Indonesia's readiness phase was supported by various arrangements and policy interventions to deal with the root causes of deforestation and forest degradation across the forest landscape, including peatlands. A number of regulations were developed to facilitate REDD+ implementation, including the creation of a national reference emission level and measurement/monitoring, reporting and verification mechanisms, as well as institutional capacity building around a national registry system, funding mechanisms, benefit and responsibility distribution, capacity building, intercommunication and coordination (Ministry of Forestry, 2010). Various models were developed before full implementation as part of the "lessons learnt" approach of the readiness phase and were undertaken in 70 "demonstration activity" areas (REDD+ Task Force, 2012a).

REDD+ has been estimated as being capable of bringing financial incentives of around US$5.6 billion to Indonesia from various international and intergovernmental sources (Busch et al., 2010). In 2008, Bappenas, the Indonesian National Development Planning Agency, claimed that around US$2.8 million was provided by the USA for protecting biodiversity and combating climate change; US$30 million was also pledged by the German Government. In addition, Australia offered A$30 million for forestry projects in Kalimantan while France and Japan pledged soft loans of US$200 million and US$400 million, respectively, for climate change mitigation actions (The Jakarta Post, 24 Oct 2008). The most prestigious grant pledge was roughly US$1 billion from the Norwegian Government in 2010 for a two-stage forest conservation programme. Stage one was planned to run from 2010 to 2014 and consisted of approximately US$200 million for preparation activities. Stage two was aimed primarily at performance-based results. It was hoped that the size of the proposed grant would catalyse other contributions (Laurance & Venter, 2010). How much of the fund has been spent, including the payment mechanism for REDD implementation, is poorly understood. Even though the government has set up a climate change trust fund to manage all funds from donors, much of the assistance goes directly to REDD+ initiatives on the ground, which makes it difficult to control and monitor. Norway's Minister for Climate and Environment complained in March 2016 that Indonesia's efforts in reducing deforestation had shown few impacts (Lang, 2016).

In order to fund various government initiatives without affecting economic stability, the government also committed to engaging the private sector and non-state actors to share the costs of emission reduction targets (mitigation) and increasing adaptive capacity (UNFCC, 2011). However, REDD+ is unlikely to get strong support from the government in terms of financial facilities. Since the forest sector now contributes significantly less to GDP than previously, the government has been more focused on other land uses activities to generate domestic income, such as palm oil plantation and the pulp and paper industry (Luttrell, Resosudarmo, Muharrom, Brockhaus, & Seymour, 2014). This situation has been exacerbated by the collapse in emissions trading markets and a commensurate decrease in demand, making other commodities, such as palm oil, more attractive (Seymour, Birdsall, & William, 2015).

Indonesia was one of the first developing countries to officially declare its commitment to voluntarily reducing emissions: in 2009 it had a target of 26% below business-as-usual predictions by 2020, and in 2015 the target was 29% by 2030 according to country's commitment to the Paris Agreement, based on unilateral actions. This target will increase a further 15% if there is adequate international support. This declaration was followed up by a number of domestic policy initiatives, including a presidential regulation (No 61/2011) for a national action plan for greenhouse gas emission reductions and a greenhouse gas inventory (No 71/2011). The Forestry Minister also created a regulation for the development of technical mechanism of REDD+ (No P.30/2009). These regulations have the potential to establish REDD+ as an effective tool for emissions reduction efforts as well as green economic development. However, there is as yet no final clarity on the regulations for REDD activities in Indonesia or fiscal incentives (Luttrell et al., 2014).[1]

Formulating and implementing REDD+ architecture in Indonesia has involved various actors at both local and national levels under an increasingly decentralised, bottom-up governance regime (Barr et al., 2006; Colfer, Dahal, & Capistrano, 2008; Moeliono, Wollenberg, & Limberg, 2009; Nurtjahjawilasa et al., 2013). But the institutional arrangements have been unstable. Three stages of REDD+ institutional formation can be seen over the period 2010–2013. The first, a task force, was an ad hoc body formed through presidential decree (No 19/2010) as a consequence of the Norway–Indonesia partnership and was focused on the establishment of a REDD+ agency establishment. The task force released a draft national strategy for REDD+ implementation and, following a presidential instruction placing a moratorium on natural forest and peat land exploitation (No 10/2011), declared Central Kalimantan Province as the first REDD+ pilot

---

[1] At the time of writing (November 2016), there were still no clear regulations about benefit-sharing arrangements for the carbon emissions reductions from REDD+ projects (how much to the donor country, how much for achieving domestic targets, and so forth), while fiscal incentives specifically for REDD+ initiatives were not yet formulated.

project area in Indonesia. This was replaced through presidential decree (No 25/2011) by a second task force comprising government and nongovernment representatives, with the intention of establishing the proposed REDD+ agency as well as implementing the national strategy and working in a transparent and accountable way. It created a draft presidential decree for the proposed agency, and strategies for financing emissions reductions initiatives and monitoring, reporting and verification. It also provided guidance on a simplified licencing system for forest law enforcement processes. This second task force successfully delineated national REDD+ governance systems and strategies as well as issues around technical implementation. At the beginning of 2013, it was replaced by a third task force under presidential decree (No 25/2011) to oversee the agency's establishment. The subsequent REDD+ agency was constituted under presidential regulation (No 62/2013) to work directly under presidential control, on a par with other ministries, to coordinate, facilitate and supervise REDD+ implementation. Subsequent political developments have significantly affected REDD+ as a national institution. By presidential regulation (No 16/2015), the newly elected president disbanded the REDD+ agency, merging it into a reformulated Ministry of Environment and Forestry (MoEF), while the ministerial regulation (No 18/2015) appointed the General Directorate of Climate Change as the national focal point for the country's emissions reduction programme (MoEF, 2015).

11.2.3 Fiscal Instruments in Indonesia

Fiscal instruments in the forestry sector refer to charges and fees imposed by government at the national level to generate revenue. They include nontaxed sources of income, such as the Forest Restoration Fund, forest licence fees and the Forest Resource Provision, and conventional taxes such as the land and building tax, value-added tax and income tax. Incentives to reduce deforestation and forest-based emissions include nonfinancial incentives such as clarification of land tenure or granting clear rights over use of the land. Financial incentives include upfront payments, such as grants, loans and investments, or results-based payments, such as payments for environmental services (which could include carbon). Other relevant instruments include the removal of perverse incentives such as increasing the accessibility of forests for activities leading to deforestation or forest degradation (which could include forest conversion).

Formal tariffs constitute the fiscal instrument most commonly used in forest management in Indonesia and follow state regulations. Forestry fiscal instruments are largely embodied in the form of non-tax state revenue (NTSR), or Penerimaan Negara Bukan Pajak, and are classified as charges or fees. Tariff policies in the forestry sector are intended to ensure the sustainability and preservation of the forest. Forestry businesses are also taxed on some conventional taxes such as land and buildings tax, income tax and value-added tax (Nurrochmat, Solihin, Ekayani, &

Hadianto, 2010). There are approximately 30 types of NTSR in the Ministry of Forestry and Environment, based on government regulation No. 12 of 2014 covering the type of NTSR tariff overseen by the ministry.

Forestry fiscal instrument policies are formulated by the central government and applied nationwide. NTSR from the forestry sector can be derived from: (1) levies imposed on the licence holders of forest products and forest area utilisation, either for rehabilitation or replacement of the forest's intrinsic value, including the Reforestation Fund and Forest Resource Provision; (2) business licence fees in forestry concession areas such as the forest licence fee; (3) fees on timber and nontimber forest products and environmental services; (4) levies from forest area utilisation for nonforest purposes; (4) fines; (5) fees for ecotourism utilisation; (6) levies imposed on utilisation of water and energy; and (7) levies on infrastructure and forestry services provision. Currently, NTSR still relies primarily on the Reforestation Fund (DR), Resource Provision (PSDH), also known as the Forest Resource Provision, and the Forest Utilisation Permit Fee (IIUPH),or the Forest Licence Fee (Mumbunan & Wahyudi, 2016; Nurfatriani, Darusman, Nurrochmat, Yustika, & Muttaqien, 2015).

In other words, the sources of NTSR from forestry can be categorised into those derived from extractive and nonextractive forestry businesses and levies on forestry-related services and infrastructure (Nurfatriani, 2016). In addition, there are fees for nonforestry-related purposes, including the substitute stumpage value (PNT), stumpage compensation (GRT) and transaction charges relating to carbon sequestration and stocking. Levies are charged to the licence holders of forestry-related business activities, including cooperatives, private-owned enterprises, state-owned enterprises and village-owned enterprises. In addition, there are further charges imposed on nonforest companies that have an impact on forest-related activities, the principal one being palm oil. A summary of these formal charges and taxes is shown in Table 11.1.

Unofficial fees, in addition to formal charges and taxes, are a further imposition on forest-related activities. They may take the form of rent capture (appropriation of charges by other actors) or bribes and usually occur at the subnational (district, local) levels; they have been estimated to cost governments across Southeast Asia billions of dollars a year (Brown, 2001). Alternatively, they may be stipulated within regulations, but not formalised in any way. Some examples, identified during the course of the research, are identified in Table 11.2.

There is a gap in administration of forestry charges and fees (Nurfatriani et al., 2015). Research reveals PSDH revenues represent only 52% of their potential (Mumbunan & Wahyudi, 2016). NTSR from the forestry sector therefore remains suboptimal (Barr et al., 2006; Barr, Dermawan, Purnomo, & Komarudin, 2010).As a consequence of decentralisation policies in Indonesia, fiscal transfers between central government and regional government have become common across the commodity sectors, and there is also a shared revenue mechanism in the forestry sector. This mechanism is only regulated on the basis of the level of timber production; hence, revenue sharing rewards producing districts based on their performance in collecting fees [DR: Dana Reboisasi (Reforestation Fund), PSDH: Provisi Sumber Daya Hutan (Indonesian: Provision of Forest Resources; Indonesia)

11 Incentives and Disincentives for Reducing Emissions under REDD+ in Indonesia

Table 11.1 Formal national charges and taxes relating to forest-related activities in Indonesia

Name	Type	Sector	Description and purpose
Provisi Sumber Daya Hutan (PSDH) or Forest Resource Provision	Charge	a. Licence holder of wood and nontimber utilisation of plantation forest, natural forest and community forest (Name of licence: IUPHHK/BK[a]) b. Licence holder of wood utilisation of ecosystem restoration in natural forest (Name of licence: IUPHHK-RE[b]) c. Licence holder of management right of village forest d. Licence holder of wood utilisation of Community Plantation Forest and Rehabilitated Plantation Forest (Name of licence: IUPHHK[c]) e. Licence holder of forest area leasing and use (Name of licence: IPPKH[d]) f. Licence holder of wood and nontimber utilisation of converted forest (Name of licence: IPK[e])	• Charges in lieu of the intrinsic value of forest products collected from the state forest and the forest areas that have been released into nonforest area or state forest reserved for nonforestry sector development • Basis for payment: Volume of forest products • Legal basic: Gov Regulation No 12/2014, MoEF regulation No 71/2016
Dana Reboisasi (DR) or Reforestation Fund	Charge	a. Licence holder of wood utilisation of natural forest (Name of licence: IUPHHK, See footnote 1) b. Licence holder of wood harvesting of natural forest (Name of licence: IPHHK[f]) c. Licence holder of wood utilisation of plantation forest from natural forest area (Name of licence: IUPHHK, See footnote 1) d. Licence holder of wood utilisation of ecosystem restoration in natural forest (Name of licence: IUPHHK-RE[a]) e. Licence holder of management right of village forest f. Licence holder of forest area leasing and use (Name of licence: IPPKH[c]) g. Licence holder of wood utilisation of Rehabilitated Plantation Forest (Name of licence: IUPHHK, See footnote 1) h. Licence holder of wood utilisation of converted forest (Name of licence: IPK[d])	• Funds for reforestation, forest rehabilitation and their supporting activities • Basis for payment: wood volume • Legal basic: Gov Regulation No 12/2014, MoEF regulation No 71/2016

(continued)

Table 11.1 (continued)

Name	Type	Sector	Description and purpose
Iuran Izin Usaha Pemanfaatan Hutan (IIUPH) or Forest Licence Fee	Charge	a. Licence holder of wood utilisation of natural and plantation forest (Name of licence: IUPHHK, See footnote 1) b. Licence holder of non-wood utilisation (Name of licence: IUPHHBK[g]) c. Licence holder of forest area utilisation for silvopastural and silvofishery system (Name of licence: IUPK[h]) d. Licence holder of environmental services utilisation in production forest (Name of licence: IUPJL[i]) g. Licence holder of wood utilisation of ecosystem restoration (Name of licence: IUPHHK-RE[a]) h. Licence holder of wood utilisation of Community Plantation Forest, Community Forest, Village Forest (Name of licence: IUPHHK, See footnote 1)	• Levies imposed on licence holders of forest utilisation in production forest on a particular forest area that being paid once at the time of the permit to be issued • Basis for payment: forest area • Legal basic: Gov Regulation No 12/2014, MoEF regulation No 71/2016

PNBP Penggunaan Kawasan Hutan or Non-tax state revenue of forest area utilisation	Charge	Licence holder of forest area leasing and use (Name of licence: IPPKH[c])	• This charge is derived from the licence of Ijin Pinjam Pakai Kawasan Hutan (IPPKH) or Licence of Forest Area Leasing and Use. This licence is for forest area utilisation for development of nonforestry purposes in the forest area over 30% of the total watershed, island and province. This imposed on the entire leased and used forest area.
			• Legal basic:
			  – Gov Regulation No 33/2014[j]
			  – MoEF regulation No 50/2016[k]

Denda Pelanggaran Eksploitasi Hutan or fine upon violation on forest exploitation		All forest concession holders	• Charge imposed due to violation on forest exploitation • Legal basic: MoEF regulation No 71/2016

Ganti Rugi Tegakan (GRT) or Stumpage compensation	Charge	All licence holders	• Levies in lieu of damaged or lost stumpage as a result of law violation actions • Legal basic: Gov Regulation No 12/2014, MoEF regulation No 71/2016
Levy of carbon sequestration and stocking transaction	Charge	Licence holder of carbon sequestration and stocking activity	• Levies imposed by 10% of the value of carbon trading • Legal basic: Gov Regulation No 12/2014
Pajak Bumi Bangunan (PBB) or Land and Building Tax in Forestry Sector	Tax	Land owner including licence holder of forest area utilisation	• Taxes imposed on land and/or buildings used for forestry business activities that granted concessions. Tax objects in the forestry sector consist of productive area, area that has not been productive yet, emplacement area, and other areas • Legal basis: – Directorate General of Tax Decree No PER—42/PJ/2015 year of 2015[l] – Circular letter of Tax Directorate General No SE-05/PJ/2016
Pajak Penghasilan (PPh) or Income Tax	Tax	Company/Enterprise, individual	• Tax charged to individual, companies or other business entities on their income • Legal basic: Law No 7/1983[m], then being revised consecutively into Law No 7/1991, Law No 10/1994, Law No 17/2000 and Law No 36/2008.
Pajak Pertambahan Nilai (PPn) or Value-Added Tax	Tax	The seller (producer, entrepreneur) who has been confirmed as the Taxable Entrepreneur (PKP)	• Tax to be imposed on every process of production and distribution, however the amount of tax charged to final consumers • Legal basic: Law No 8/1983[n] then being revised by Law No 42/2009

(continued)

Table 11.1 (continued)

Name	Type	Sector	Description and purpose
Levies on exports of Oil Palm commodities and its derivatives—contributes to crude palm oil (CPO) levy for the Indonesian Oil Palm Estate Fund—BPDP)	Charges	a. Palm oil plantation company, which exports palm oil commodities and its derivatives b. Raw material industry of palm oil c. Exporter of oil palm-related commodities and its derivatives	• Fees on the export of palm oil plantation commodities and its derivative • Legal basic: – Gov Regulation No 24/2015[o] – Gov Regulation No 61/2015[p] – MoF regulation No 133/2015[q]
Fee on palm oil plantation companies – contributes to crude palm oil (CPO) levy for the Indonesian Oil Palm Estate Fund—BPDP)	Charges	Palm oil plantation company	• Fees collected for the sustainable palm oil development • Legal basic: – Gov Regulation No 24/2015 – Gov Regulation No 61/2015 – MoF regulation No 133/2015

[a]Izin Usaha Pemanfaatan Hasil Hutan Kayu/Bukan Kayu (Licence of wood and nontimber utilisation)
[b]Izin Usaha Pemanfaatan Hasil Hutan Kayu Restorasi Ekosistem (Licence of wood utilisation of ecosystem restoration)
[c]Izin Usaha Pemanfaatan Hasil Hutan Kayu (Licence of wood utilisation)
[d]Izin Pinjam Pakai Kawasan Hutan (Licence of forest area leasing and use)
[e]Ijin Pemanfaatan Kayu (Licence of wood utilisation of converted forest)
[f]Izin Pemungutan Hasil Hutan Kayu (Licence of wood harvesting)
[g]Izin Usaha Pemanfaatan Hasil Hutan Bukan Kayu (Licence of nontimber utilisation)
[h]Izin Usaha Pemanfaatan Kawasan (Licence of forest area utilisation)
[i]Izin Usaha Pemanfaatan Jasa Lingkungan (Licence of environmental services utilisation)
[j]MoFor (2015). Government Regulation No 33/2014 regarding Type and Rate of Non-Tax State Revenue from Forest. Jakarta
[k]MoEF, Ministry of Environment and Forestry Government of Indonesia (2016). Decree No 50/2016 regarding Guidance of Forest Area Lease and Use. Jakarta
[l]Director General of Tax Decree No PER - 42/PJ/2015 regarding Procedures of Land and Building Tax Imposition for Forestry Sector

[m]MoF (1983). Law No 7/1983 regarding Income Tax, Jakarta
[n]MoF (1983). Law No 8/1983 then being revised by Law No 42/2009 regarding value-added tax for goods and services and sales tax on luxury goods, Jakarta Ministry of Finance, Government of Indonesia (2009). Law No 8/1983 then being revised by Law No 42/2009 regarding value-added tax for goods and services and sales tax on luxury goods
[o]MoEF. (2015). Government Regulation No 24/2015 regarding Plantation Fund Collection. Jakarta
[p]MoF. (2015). Government Regulation No. 61/2015 regarding CPO Fund Collection and Utilization. Jakarta
[q]MoF. (2015). Decree No. 133/2015 regarding Rate of General Service Agency CPO Fund Agency in Ministry of Finance Ministry of Forestry, MoFor (2010). Strategi REDD-Indonesia Fase Readiness 2009—2012 dan progress implementasinya. M. o. Forestry. Jakarta

Table 11.2 Selection of informal subnational charges relating to forest-related activities in Indonesia

Title	Type	Sector	Description and purpose
Stumpage fees for local community	Unofficial charge	Native forest	Varied for each company depending on agreement between company and community (ranging from 150,000 Indonesian rupiah to 35,000 per M^3) Propose: as benefit sharing of natural resources to the local community
Log transportation fees from forest to log aggregation point	Unofficial charge to local police	Native forest	To get unofficial permission from police to transport logs (by road or river) from the forest to the log aggregation point
Community development charge	Stated in the decree of the concession but not clarified how much should be allocated	Native and plantation forest	To develop local community livelihood
Donation to district government	Unofficial charge	Native and plantation forest	To support an event held by district government

arising from timber production activities (Nurfatriani et al., 2015). Charges for using forest areas for nonforest purposes are not expected to increase over time, since the more revenue that comes from this charge, the more forest is converted (Prayitno, Taufik, Fitriyani, Ramdan, & Gunawan, 2013). The CPO fund, introduced in 2015, is intended for food security, encouraging the palm oil plantation industry to relocate downstream from other agricultural activities, as well as supporting the supply and use of biodiesel. The government established the CPO Fund Management Agency under the Finance Ministry to collect, administer, manage, store and distribute the CPO fund. According to the agency, the CPO fund will amount to 11 trillion Indonesian rupiah in 2016. It will be channelled to support government programmes that mix biodiesel in diesel fuel where currently the composition of biodiesel is only 20%. The aim is to improve the quality of palm oil production and empower farmers. However, there is a concern that this will trigger an extension of oil palm plantations and further promote the expansion of forestland occupation and encroachment by local communities and others for oil palm plantations. The fund also explicitly aims to support the replanting of up to 100,000 hectares of plantations in 2016 (Anonymous, 2015; Purnomo et al., 2016; UNEP, 2016). Therefore, the use of CPO funds to support the development of biodiesel should be aligned with the interests of reducing forest conversion for oil palm plantation in the context of REDD+.

11.3 Conclusions

Indonesian fiscal policy has provided a monetary incentive for local government, especially forest-rich districts, to exploit their timber resources (Nurrochmat, 2005; Nurrochmat et al., 2010). This situation is considered as a disincentive for REDD+ implementation in Indonesia. In order to avoid the raising of revenue as a driver of degradation and deforestation, it has been suggested that improved financial governance is required to encourage policy reform. Reforms should include the removal of perverse incentives for forest conversion, the equitable and accountable distribution of financial incentives, the prevention of corruption and fraud, and the strengthening of economic benefits for smallholders (Barr & Sayer, 2012). REDD+ activities themselves have the potential to act as a disincentive, as they include the costs of the measurement of carbon sequestration and carbon stock in the area allocated for REDD+ activity. In addition, the levy of 10% of the transaction value for carbon sequestration and stocking as imposed now represents a disincentive for REDD+ activity (Nurfatriani, 2016). To avoid these impositions, REDD+ payments should compensate, at the very least, for the costs of implementation, including the opportunity costs to local government, the community and the forest private sector. To ensure support for REDD+, payments need to be designed to exceed stakeholders' forgone revenues from land use change (Irawan, Tacconi, & Ring, 2013; Ring, Drechsler, van Teeffelen, Irawan, & Venter, 2010).

The Government of Indonesia needs to consider providing incentives for the nonexploitation of forests by businesses engaged in the provision of environmental services as well as carbon transactions. This could take the form of private investments, private–public partnerships or civil society engagement in forestry and land use change, and may include incentives such as payment for ecosystem services and forest ecosystem restoration. Specifically in the forest and land use sector, the government aims to encourage investments in sustainable forest management involving policies that put higher value on sustainable forests through economic instruments (for example, royalties and fees); enforcement (control of illegal logging, forest fires and forest land conversion); and policies affecting demand (for example, expanding the use of wood certification and accreditation systems and benefits associated with corporate social responsibility); as well as reducing risks associated with unclear land ownership. This will also thus entail addressing institutional issues related to governance and capacity building in all government levels, given Indonesia's high degree of decentralisation, to ensure effectiveness.

References

Angelsen, A., & McNeill, D. (2012). The evolution of REDD. In A. Angelsen, M. Brockhaus, W. D. Sunderlin, & L. V. Verchot (Eds.), *Analysing REDD: Challenges and choices* (pp. 31–48). Bogor, Indonesia: CIFOR.

Anonymous. (2015, September 7). Disbursement of CPO funds ready to spur industry. *The Jakarta Post*.

Barr, C., Dermawan, A., Purnomo, H., & Komarudin, H. (2010). *Financial governance and Indonesia's Reforestation Fund during the Soeharto and post-Soeharto periods, 1989–2009: A political economic analysis of lessons for REDD*. Bogor, Indonesia: CIFOR.

Barr, C., Resosudarmo, I. A. P., Dermawan, A., McCarthy, J., Moeliono, M., & Setiono, B. (2006). *Decentralization of forest administration in Indonesia: Implications for forest sustainability, economic development and community livelihood*. Bogor, Indonesia: CIFOR.

Barr, C. M., & Sayer, J. A. (2012). The political economy of reforestation and forest restoration in Asia–Pacific: Critical issues for REDD. *Biological Conservation, 154,* 9–19.

Brown, D. W. (2001). *Why governments fail to capture economic rent: the unofficial appropriation of rain forest rent by rulers in insular southeast Asia between 1970 and 1999*. (PhD thesis), University of Washington, USA.

Busch, J., Strassburg, B., Cattaneo, A., Lubowski, R., Bruner, A., Rice, R., … Boltz, F. (2010). Open source impacts of REDD incentives spreadsheet (OSIRIS v3.4). *Collaborative Modeling Initiative on REDD Economics*. September 2010.

Chapman, S., Wilder, M., & Millar, I. (2014). Defining the legal elements of benefit sharing in the context of REDD. *Carbon & Climate Law Review, 8*(4), 270–281.

Chokkalingam, U., & Vanniarachchy, A. (2011). *Sri Lanka's REDD potential, myth or reality?* Colombo: Forest Carbon Asia.

Colfer, C. J. P., Dahal, G. R., & Capistrano, D. (2008). *Lessons from forest decentralization: Money, justice and the quest for good governance in Asia-Pacific*. London: Earthscan.

Eliasch, J. (2008). *Climate change: Financing global forests: the Eliasch Review*. London: Earthscan.

Henderson, I., Coello, J., Fischer, R., Mulder, I., & Christophersen, T. (2013). The role of the private sector in REDD: The case for engagement and options for intervention, UN-REDD Program. *Policy Brief, 4,* 1–12.

ICIMOD (The International Centre for Integrated Mountain Development). (2016). *Regional workshop on 'REDD safeguards for Himalayas*. Retrieved from http://www.icimod.org/?q=23022.

Irawan, S., Tacconi, L., & Ring, I. (2013). Stakeholders' incentives for land-use change and REDD+: The case of Indonesia. *Ecological Economics, 87,* 75–83.

Kanninen, M., Murdiyarso, D., Seymour, F., Angelsen, A., Wunder, S., & German, L. (2007). *Do trees grow on money? The implications of deforestation research for policies to promote REDD*. Bogor: Center for International Forestry Research.

Lang, C. (2016). *Norway admits that 'We haven't seen actual progress in reducing deforestation' in Indonesia*. Retrieved from http://www.redd-monitor.org/2016/03/03/norway-admits-that-we-havent-seen-actual-progress-in-reducing-deforestation-in-indonesia/#more-22668

Laurance, W. F., & Venter, O. (2010). Measuring forest changes. *Science, 328*(5978), 569–569.

Lubell, M. (2015). Collaborative partnerships in complex institutional systems. *Current Opinion in Environmental Sustainability, 12,* 41–47.

Luttrell, C., Resosudarmo, I. A. P., Muharrom, E., Brockhaus, M., & Seymour, F. (2014). The political context of REDD in Indonesia: Constituencies for change. *Environmental Science & Policy, 35,* 67–75.

McAfee, K. (2016). Green economy and carbon market for conservation and development: A critical view. International environmental agreements. *Politics, Law and Economics, 16*(3), 333–353.

Ministry of Finance, MoF. (2012). *Indonesia's first mitigation fiscal framework*. Retrieved from https://www.unpei.org/sites/default/files/e_library_documents/Indonesia_MFF_report.pdf

Ministry of Forestry. (2010). Strategi REDD-Indonesia Fase Readiness 2009–2012 dan progress implementasinya. Ministry of Forestry, Jakarta.

MoEF. (2015). *Regulation No 18, 2015 on the organization and working procedure of the Ministry of Environment and Forestry*.

Moeliono, M., Wollenberg, E., & Limberg, G. (2009). *The decentralization of forest governance: Politics, economics and the fight for control of forests in Indonesian Borneo*. London: Earthscan.

Mumbunan, S., & Wahyudi, R. (2016). Revenue loss from legal timber in Indonesia. *Forest Policy and Economics, 71*(C), 115–123.

Nurfatriani, F. (2016). Formulasi Kerangka dan Strategi Implementasi Kebijakan Fiskal Pembangunan RendahKarbon di SektorKehutanan (dissertation) Bogor, ID: InstitutPertanian Bogor.

Nurfatriani, F., Darusman, D., Nurrochmat, D. R., Yustika, A. E., & Muttaqien, M. Z. (2015). Redesigning Indonesian forest fiscal policy to support forest conservation. *Forest Policy and Economics, 61*, 39–50.

Nurrochmat, D. R. (2005). *The impacts of regional autonomy on political dynamics, socio-economics and forest degradation. Case of Jambi Indonesia* (Dissertation). Goettingen: University of Goettingen.

Nurrochmat, D. R., Solihin, I., Ekayani, M., & Hadianto, A. (2010). *Neraca Pembangunan Hijau-Konsep dan Implikasi Bisni sKarbondan Tata Air di Sektor Kehutanan*. Bogor, ID: IPB Press.

Nurtjahjawilasa, D., Ysman, K., Septiani, I., & Lasmini, Y. (2013). *Modul: Konsep REDD dan implementasinya*. Jakarta: The Nature Conservancy.

Palmujoki, E., & Virtanen, P. (2016). Global, National, or Market? Emerging REDD governance practices in Mozambique and Tanzania. *Global Environmental Politics, 16*(1), 59–78.

Pan, Y., Birdsey, R. A., Fang, J., Houghton, R., Kauppi, P. E., Kurz, W. A., et al. (2011). A large and persistent carbon sink in the world's forests. *Science, 333*(6045), 988–993.

Prayitno, H., Taufik, A., Fitriyani, R., Ramdan, D., & Gunawan, P. R. A. S. (2013). *Mengukur komitmen: Analisis kebijakan perencanaan dan anggaran nasionalterhadap pengelolaan hutandan lahan di Indonesia*. Jakarta: FITRA.

Purnomo, H., Dewayani, A. A., Achdiawan, R., Ali, M., Komar, S., Okarda, B., & Juniwaty, K. S. (2016). *Tata kelola rantainilaisawitdankebakaranhutandanlahan*. Makalahpada FGD Penguatan.

REDD Task Force (BadanPengelola REDD). (2012a). BP-REDD Jelaskan Status Demonstration Activities di Indonesia. Retrieved from http://bpredd.reddplusid.org/component/content/article?id=1682:bp-redd-jelaskan-status-demonstration-activities-di-indonesia

REDD Task Force. (2012b). *REDD national strategy*. Bogor: Indonesian REDD Task force.

Righelato, R., & Spracklen, D. V. (2007). Environment. Carbon mitigation by biofuels or by saving and restoring forests? *Science, 317*(5840), 902. https://doi.org/10.1126/science.1141361.

Ring, I., Drechsler, M., van Teeffelen, A. J. A., Irawan, S., & Venter, O. (2010). Biodiversity conservation and climate mitigation: what role can economic instruments play? *Current Opinion in Environmental Sustainability., 2*, 50–58.

Sahide, M. A. K., & Giessen, L. (2015). The fragmented land use administration in Indonesia—Analysing bureaucratic responsibilities influencing tropical rainforest transformation systems. *Land Use Policy, 43*, 96–110.

Seymour, F., Birdsall, N., & William, S. (2015). *The Indonesia-Norway REDD+ agreement: A glass half-full. CGD policy paper 56*. Washington, DC: Centre for Global Development.

Somorin, O. A., Visseren-Hamakers, I. J., Arts, B., Sonwa, D. J., & Tiani, A. M. (2014). REDD policy strategy in Cameroon: Actors, institutions and governance. *Environmental Science & Policy, 35*, 87–97.

Stern, N. (2007). *The economics of climate change. The Stern review*. Cambridge: Cambridge University Press.

UNEP. (2016). *Fiscal incentives for Indonesian palm oil production: Pathways for alignment with green growth*. United Nations Development Programme: UN-REDD Programme.

UNFCCC. (2011). *Outcome of the work of the ad hoc working group on long-term cooperative action under the convention*. United Nations, New York: Cancun Agreements.

Venter, O., & Koh, L. P. (2012). Reducing emissions from deforestation and forest degradation (REDD): Game changer or just another quick fix?(Report). *Annals of the New York Academy of Sciences, 1249*, 137–150.

Vijge, M. J., Brockhaus, M., Di Gregorio, M., & Muharrom, E. (2016). Framing national REDD benefits, monitoring, governance and finance: A comparative analysis of seven countries. *Global Environmental Change, 39*, 57–68.

World Bank. (2015). *World Bank Board committee authorizes release of revised draft environmental and social framework*. Retrieved from http://www.worldbank.org/en/news/press-release/2015/08/04/world-bank-board-committee-authorizes-release-of-revised-draft-environmental-and-social-framework

Chapter 12
Carbon Budgeting Post-COP21: The Need for an Equitable Strategy for Meeting CO_2e Targets

Robert Hales and Brendan Mackey

Abstract The Paris Climate Change Agreement is a mixed blessing. Although it has been heralded as a great success, there is a low probability that it will reduce carbon dioxide equivalent emissions at the pace required to ensure a safe climate, based on current progress and processes. This chapter provides an overview of the scale of the problem of achieving the Paris Climate Change aspirations using a carbon emissions budget approach. This is important because current National Determined Contributions and Intended National Determined Contributions do not place the global community on a pathway to limit warming at or below 2 °C above pre-industrial levels by 2100. Suggestions are made on how a general carbon emissions budget can be used to limit global warming through equity and transparency processes. This chapter adds to the conversation on the imperative to ratchet up commitments under the Paris Agreement as the window on action to mitigate against dangerous climate change is about to close.

Keywords Climate change • Paris agreement • COP21 • Net zero emissions • CO_2 budget • Stocktake mechanism

12.1 Introduction

Despite the achievements of the Paris Agreement at COP21, the world is still heading towards a climatic disaster in the short to medium term (Hansen et al., 2017). Current climate mitigation efforts are too small, too limited and too slow to achieve the ambitious targets of the Paris Agreement (Peters et al., 2017). To prevent catastrophic climate change, the global economy will have to be rapidly restructured to

---

R. Hales (✉)
Griffith Centre for Sustainable Enterprise, Griffith Business School, Griffith University, Nathan, Queensland, Australia
e-mail: r.hales@griffith.edu.au

B. Mackey
Griffith Climate Change Response Program, Griffith University, Southport, QLD, Australia

ensure the global economy decouples economic growth from carbon dioxide equivalent (CO_2e) emissions (Dietz & O'Neill, 2013; du Pont, Jeffery, Gutschow, Christoff, & Meinshausen, 2016). This will require a massive response similar in scale and urgency to the Allied Forces' effort in the aftermath of the Second World War to rebuild Europe and the world economy. However, compared to the aftermath of that war, there appears to be little appetite for the strong, appropriate and coordinated international action needed to combat climate change. There are some signs that CO_2e emissions can be decoupled from economic growth but the scale and pace of change needed is far greater than present trends (Schandl et al., 2016).

The COP21 negotiations were limited from the outset by the need to work within existing political, economic and technological frameworks. Because governments with differing priorities and interests tend to agree on minimal targets tempered by short-term national self-interest, most responses have been reactive rather than proactive, incremental, partial and voluntary, with little concrete action that would lead to deep and rapid cuts in emissions and systemic change. Existing strategies are unlikely to produce the radical transformations required to rapidly end carbon pollution: for example, no comprehensive plans have been developed for making the transition to a sustainable global system (Althor, Watson, & Fuller, 2016). Furthermore, the exit (and renegotiation of participation) of the USA from the Paris Agreement under the Trump administration demonstrates that even if nation states agree to the targets and align their policies and actions within the Paris Agreement, they can exit the Agreement if they wish. These factors heighten the need for all nations to ensure that the national targets align with policies and economic activities to limit warming to well below 2 °C above pre-industrial levels by 2100.

The Paris Agreement has been rightly hailed as a success in, among other things, cementing the below 2 °C goal and going further to reference 1.5 °C and have more nuanced differentiation between developed and developing countries (Kinley, 2017). However, the Agreement's compliance mechanisms are weak, being limited to a regular global stocktake that will inform national determined responses and a mechanism that will operate in a nonpunitive manner. An alternative approach is needed: one that starts by determining what is necessary to achieve safe outcomes and then back casts to design and develop viable solutions. This "critical-safety" approach is widely used in industry to manage critical, complex projects. However, in the current economic and political frameworks, this approach is not being adopted by governments to manage climate risk because in order to preserve critical ecosystems (and the ecosystem service benefits they provide), strict limits on both pollution and consumption need to be implemented. This dichotomy—between the need to work within the Paris Agreement's rules, processes and institutions, and the reality that they are structurally weak—needs to be addressed quickly.

Global carbon emissions budget approaches have been previously proposed to ensure that equity of emissions reductions is aligned with the purpose of the Paris Agreement. These proposals range from equity considerations and GDP (Wang & Watson, 2012), allocation at a country level using per capita carbon quotas within a global carbon budget and contractions to an agreed year at which per capita emissions would converge to a global average (Meyer, 2000), combinations of historical emissions, quotas and ability to pay (Mattoo & Subramanian, 2012), multifactor equity principles, and per capita calculation of obligations under an emission cap (Robiou et al., 2017). The purpose in this chapter is not to argue for a particular method, but rather to lay the general case for a global carbon emissions budget approach that ensures the Nationally Determined Contributions (NDCs) limit warming to well below 2 °C above pre-industrial levels by 2010 as per the Paris Agreement.

The argument of this chapter is, firstly, that maximising climate mitigation efforts within current political and economic growth agendas needs to occur by increasing commitments to NDCs constrained by a global carbon emissions budget that is likely to limit global warming to well below 2 °C within the existing fair share approach. Secondly, the systemic transformation and the emergence of a global budget approach to reducing emissions may occur through the transparency processes in the Paris Agreement. The budget approach has the advantage of setting a non-arbitrary limit on global emissions from which economic growth can be decoupled from increasing emissions. Such a breakthrough is a vital part in the process of achieving a global sustainable economy in the longer term. Before outlining a fair share and transparency approach, the global CO_2e budget and the Paris Agreement will be explored to reveal the shortcomings of current national mitigation targets.

12.2 The Global CO_2e Budget and the Paris Climate Change Agreement

To keep global average temperature below 2 °C above pre-industrial levels, the world can emit 2900 billion tonnes CO_2e from 1870 to 2100 (IPCC, 2014; Le Quere et al., 2016). Since 1870, the world has emitted about 2100 billion tonnes of CO_2e. In other words, 73% of the allowable global carbon has been emitted. To keep temperatures from exceeding 2 °C above pre-industrial levels, the remaining 27% (800 billion tonnes of CO_2e) can be emitted between 2017 and 2100: the equivalent of about 22 years worth of current global annual anthropogenic emissions (~36 billion tonnes CO_2e; Le Quere et al., 2016). The difference between the past 146 years and the present to 2100 is that from now on CO_2e emissions need to decrease every year until 2100. To illustrate the problem, three figures from three organisations are presented. Not only does this graphical display of data illustrate the inherent problem in the Paris Agreement, the figures have been drawn from websites that publicise the relative responsibilities and commitments of nations and thus are an integral part of the transparency process in the negotiation of developing national mitigation plans. The sites analyse and publicise the importance of a global safety net or global budget approach.

Despite the fanfare over the Paris Agreement, the current commitments by countries (NDCs) do not set the world on track to mitigate against a major global warming scenario (Rogelj, den Elzen, et al. 2016). The current mitigation pledges by countries under the Agreement will not be enough to ensure that less than 800 billion tonnes of CO_2e are emitted by 2100. Figure 12.1 indicates the trajectory of

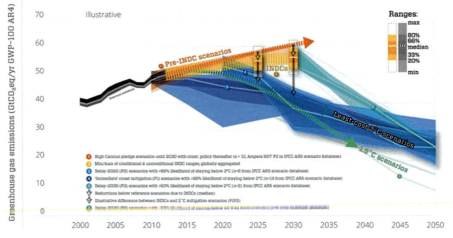

Fig. 12.1 Aggregate effect of the intended nationally determined contributions on greenhouse gas emissions. Source: UNFCCC (2016)

current pledges from Parties to the Paris Agreement. It is important to note that the current NDCs and Intended National Determined Contributions (INDCs) do not translate into the scenarios of the 1.5 °C or the 2 °C targets.

Similarly, under a budget model approach, the current NDCs and INDCs do not translate into a high likelihood of achieving a 2 °C target. Figure 12.2 compares the present NDCs and their greenhouse gas emissions and a sample (i.e. indicative) pathways to reach the 2 °C target (Global Carbon Budget, 2016).

Figure 12.2 illustrates that the emissions budget of the four largest emitters under the NDCs scenarios exceeds the budget (at 2030) to limit warming to less than 2 °C below pre-industrial levels. At this point, even if the rest of the world planned to make radical changes to their NDCs, their actions would not help achieve the Paris Agreement target of limiting global warming to less than 2 °C below pre-industrial levels. Figure 12.3 also illustrates that under the national plans (NDCs), emissions will continue to rise and exceed the budget of allowable emissions (Climate Interactive, 2016).

Figure 12.3 illustrates that the current national plans (NDCs) will not limit global warming to 2 °C. These points are further illustrated by the carbon tracker publication of national commitments and resultant emissions trajectories (Carbon Tracker, 2016).

Consistent with other assessments of global budgeting of greenhouse emissions and the target of limiting warming at or below 2 °C (Rogelj, Schaeffer, et al. 2016), the Carbon Tracker (2016) indicates that current national policies will not achieve the ambition of the Paris Agreement and will only limit global warming to 3.3–3.9 °C

12 Carbon Budgeting Post-COP21: The Need for an Equitable Strategy for Meeting... 213

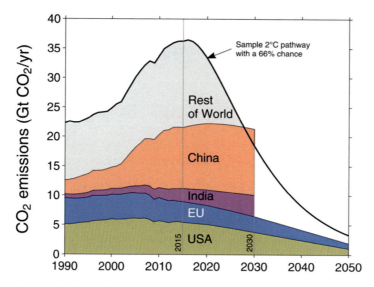

Fig. 12.2 Current NDCs and the exceedance of emissions under a sample 2 °C pathway. Source: Global Carbon Budget 2016

in 2100. Furthermore, if the pledges INDCs are considered within projection scenarios, the actions of countries will only limit warming to 2.5–2.8 °C (Fig. 12.4).

Time is rapidly running out to limit warming to 2 °C and even more so to limit it to 1.5 °C (Rogelj, Schaeffer, et al. 2016). As a result, the INDCs of all countries need to be revisited and an increase in their commitment is needed to achieve a drastic reduction in emissions in line with the global carbon budget of ~800 billion tonnes CO_2e.

12.2.1 Fair Share

One of the critical roadblocks in international negotiations on climate change mitigation is the difficulty of agreeing on a country's fair share of total emissions reductions as part of meeting obligations under the Paris Agreement. The current process of determining a fair share is one where each country determines what is feasible within the political processes of their nation and the transparency of the information provided in support of their NDCs allows for evaluation by other signatories. Given that the state parties' commitments under the Paris Agreement are nationally determined and the compliance process is weak and essentially based on volunteerism, the scrutiny of national commitments by other nations is the primary mechanism for holding countries accountable to delivering their emission reduction pledges. This is unlike the previous Kyoto Protocol where punitive measures for non-compliance were part of that agreement (UNFCCC, 1997).

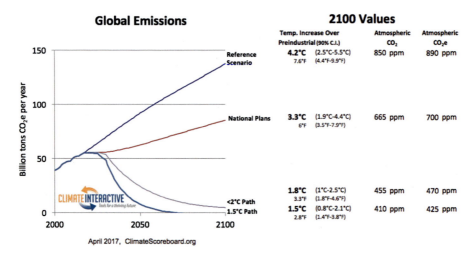

Fig. 12.3 Global CO₂e emissions trajectories under business as usual and 2016 NDCs scenarios. Source: Climate Interactive (2016)

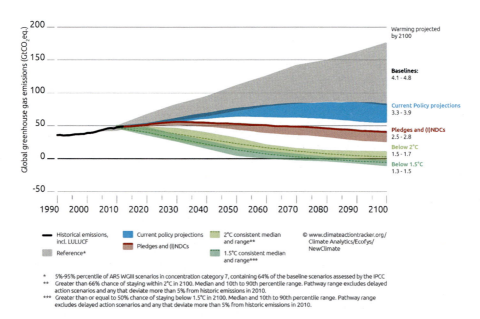

Fig. 12.4 Current (2016) pledges and INDCs do not limit global warming below 2 °C. Source: Carbon Tracker (2016)

How NDCs mitigation targets are agreed upon in each country will be governed by political debate internal to that country that inevitably reflects the tensions that arise between promotion of short-term national self-interest and advancing mitigation target that are fair in a global context, and this is where the problem lies.

There are a number of factors that need to be considered when each country determines their national contribution to the lowering of global emissions. Under the Paris Agreement process, determining the fair allocation of the global carbon budget is based upon principles of equity and differentiated responsibilities and respective capabilities. The IPCC has proposed equity principles that comprise four key dimensions: responsibility, capacity, equality and the right to sustainable development (IPCC, 2014). Each nation needs to determine its allocation of reductions as part of the global carbon budget after considering each of those dimensions:

1. Responsibility—historical responsibility for current elevated greenhouse gas atmospheric concentrations.
2. Equality—the equal right of each citizen to use the atmosphere as a sink for their greenhouse gases.
3. Capacity—the ability of a nation to reduce its greenhouse emissions.
4. Development—the right to development, particularly in the context of sustainable development and efforts to eradicate poverty.

As argued by Mackey and Rogers (2015), the per capita approach is based explicitly on the ethical principle that each human being has equal rights as stated in Article 1 of the Universal Declaration of Human Rights. It follows that each human has equal rights to the ecosystem services provided by the global commons, in this case the carbon-absorbing capacity of the Earth system. Determining historical responsibility is a contested topic. The point at which history of emissions is taken into consideration is important in the equity considerations of NDCs and INDCs (Peters, Andrew, Solomon, & Friedlingstein, 2015; Skeie et al., 2017). The large volume of historical emissions of countries that resulted from economic growth prior to the UN conventions on climate change is an important consideration if the concept of a responsibility is to be fully accounted under a fair share process. However, this is unlikely. Balancing all four considerations as per the circumstances of each country is the sovereign right of each nation state; however, internal politics appears to trump long-term collective concerns of global warming.

The steps each state party should undertake as part of a process of a fair share under a global carbon budget approach include, firstly, articulating which of the equity principles or combination of principles (i.e. 1–4 above) they have adopted in determining their national determined mitigation commitment and how these reflect their national context. Next, each state party needs to describe how they have operationalised the principles, e.g. what do they consider to be their historical responsibility. Finally, their mitigation target should be specific in terms of what portion of the global carbon budget it appropriates, in addition to specifying it in terms of achieving by a future date a percentage reduction relative to a base year. This should be revealed under the Paris Agreement's stocktake mechanism. This process of identifying the portion of the global carbon budget being appropriate by a nationally determined mitigation commitments level is not referenced explicitly in the Paris Agreement. In fact, the NDC approach was advanced as a way around the roadblock that had been struck during negotiations around such "bottom-down" approaches such as that which led to the Kyoto Protocol. The consequence, however, of not explicitly considering the global carbon budget is evidenced by the Figs. 12.1, 12.2 and 12.3 that show how the present NDCs do not lead to a 2 °C pathway.

Now that President Trump has announced the USA's intention to withdraw from Paris Agreement, there is great uncertainty that US emissions reductions will occur as per the NDC. This compounds the problem of determining a fair share of emissions reductions within the current process of negotiation. Other Party members of the Paris Accord will need to ensure their NDCs are such that they take into account the USA's potential failure to reduce emissions. This puts more pressure on all of the remaining signatories to further reduce their emissions and at a faster rate than anticipated, to compensate for the potential increase in US emissions. This specifically puts greater pressure on the largest emitters (China, India and the EU) to adopt a global carbon budget approach if the Paris Agreement is to limit warming to well below 2 °C above pre-industrial levels by 2100.

12.2.2 Transparency

The primary method to maximise climate mitigation efforts within current political and economic growth agendas is to ensure the process and outcomes of increasing NDCs are transparent (Winkler, Mantlana, & Letete, 2017). The Paris Agreement Article 13 states that "The Nationally determined contributions communicated by Parties shall be recorded in a public registry maintained by the secretariat" (UNFCCC, 2015, p. 7). Transparency of national commitment to reducing emissions under the Paris Agreement is a vital process to determine the fair share of reductions. The presence of a voluntary process of commitment to increasing national targets over time has been attractive to nation states.

Despite the acknowledgement of the importance of transparency in the Paris Agreement, the process for transparency is a matter of debate and is still evolving (Prinn & Reilly, 2017). Developing a robust transparency framework before COP24 (2018) is vital for the success of a global budget approach to the Paris Agreement. How Parties will account for anthropogenic emissions and removals corresponding to their NDCs will be guided by the promotion of "environmental integrity, transparency, accuracy, completeness, comparability and consistency, and ensure the avoidance of double counting, in accordance with guidance adopted by the Conference of the Parties serving as the meeting of the Parties to the Paris Agreement" (UNFCCC, 2015). To enable this, three approaches to transparency will be part of the process of submitting NDCs: stocktake mechanism, compliance mechanism and the expert review process.

The stocktake mechanism of the Paris Agreement allows for each nation to commit greater contributions to decrease emissions. Starting in 2018, the UNFCCC stocktake process provides all Parties with the opportunity to increase and harmonise the ambition contribution under the Common but Differentiated Responsibilities and Respective Capabilities equity principle, to determine how nations are complying with their obligations under the Paris Agreement (see UNFCCC, 1992). The Parties will describe how they have enacted policies to reduce greenhouse gas emissions and whether those policies have reduced those emissions. The stocktake mechanism also allows for an important transparency process in that developed countries document their financing obligations to developing countries.

However, the five-yearly global stocktake process starting in 2020 does not yet have a robust mechanism to ensure transparency of an equitable outcome and one that ultimately achieves the Paris goal (Briner & Moarif, 2017). The main reason for the potential failure of NDCs at the first stocktake is a disconnect between the expert reviews of national reports, the compliance mechanism and the ultimate goal of keeping the increase in the global average temperature well below 2 °C above pre-industrial levels (Stockholm Institute, 2016). Article 15 of the Paris Agreement does not clearly define the scope and functions of the mechanism. So, while expert reviews may determine that an NDC of a country falls short of its fair share (possibly using a global carbon budget approach), the compliance mechanism hinges upon pressure being exerted by Parties during negotiations surrounding the five-year stocktake mechanism deadline. The first critical millstone in this process is the 2018 ambitions deadline. Political processes and civil society within each nation can create pressure to make potential increases in commitments to NDCs at this time. The pressure on governments to increase their commitments from the ground up relies on democratic processes to generate a political imperative and provide a social licence to make greater increases. Thus, under the current process of stocktake mechanism and its accompanying transparency, there is an over-reliance on advocacy from civil society and the machinations of domestic electoral politics to ratchet up the commitments of countries that lag behind what is needed and warranted, given their national circumstances. In short, while the transparency process is one that supports the democratisation of emission reduction pledges at the country level, it offers no specific and concrete mechanism that the international community can deploy such that there are concrete consequences for those whose commitments and actions fail to be judged reasonable, appropriate and fair. We suggest that if a global carbon budget approach was explicitly used in the global stocktake process it may offer more leverage for pressure, as it provides an absolute benchmark to judge NDCs individually and collectively.

Other issues with the transparency process is that there remains a lack of clarity about reporting requirements for developing country Parties and such reporting may be burdensome. There is also an issue regarding the reporting of national emissions for developing countries and how that is linked with attracting climate adaptation and mitigation finance. There may be positive and negative finance implications resulting from reporting high or low emissions. In the Paris Agreement, transparency surrounding the reduction of emissions from deforestation and forest degradation (Article 5), cooperative mechanisms (Article 6) and loss and damage (Article 8) is not explicitly mentioned. Given the disparity between historical emissions of developed and developing countries, it is vital for transparent processes to highlight the NDCs of both developed and developing countries that have been achieved through the reduction of emissions through avoiding deforestation and degradation and increasing reafforestation.

12.3 The Budget Approach

The process of increasing commitment through the NDCs suits the political process but runs a high risk of failing because the process does not focus on the part each country can or should play in the 2 °C pathway under the global carbon budget approach. NDCs should be consistent with a global carbon budget with the aim of achieving the global warming limit at an acceptable level of probability.

There are three steps that can assist Parties to more explicitly and effectively achieve their NDCs that conform to a global budget approach within the Paris Agreement. The application of the three steps will also generate the information required to facilitate evaluation of the adequacy of an NDC's compliance with the Agreement's warming limit goal and equity requirements. Even if nations do not formally follow these steps, any national greenhouse gas emissions reduction target is implicitly a position on questions raised by these steps:

1. Include within the stocktake mechanism of the Paris Agreement the concept of a carbon emissions cap whereby the global carbon emissions budget of 800 billion tonnes of CO_2e can be negotiated by Parties between 2017 and 2100.
2. Each country determines their NDCs in line with their budgeted carbon emissions based upon equity and common but differentiated responsibilities and respective capabilities.
3. Publish INDCs that conform to the global carbon emissions budget approach as a process of negotiation prior to stocktake mechanism review. Not only are INDCs published but expert review by the UNFCCC plays a role in providing feedback on the effectiveness of achieving an emissions trajectory that conforms to a global carbon emissions budget.

Given the urgency of the task, the present procedures of the Paris Agreement may well be inappropriate for the crisis facing the globe. The future climate of Earth will be determined by the extent to which the actions of nations that reduce emissions leading up to 2030 at a greater than current ambitions (Rose, Richels, Blanford, & Rutherford, 2017). Even if the specific steps recommended here are not formally incorporated into the global stocktake process, it is critical that the concept of the global carbon budget become universally understood and that all stakeholders—especially the governments that represent the interests of ordinary citizens and future generations—realise its pivotal significance in the success of the Paris Agreement's implementation (Fig. 12.4).

References

Althor, G., Watson, J. E., & Fuller, R. A. (2016). Global mismatch between greenhouse gas emissions and the burden of climate change. *Scientific reports, 6*, srep20281.

Briner, G., & Moarif, S. (2017). Possible structure of mitigation-related modalities, procedures and guidelines for the enhanced transparency framework. *OECD/IEA Climate Change Expert Group Papers, 2016/05*, 18. https://doi.org/10.1787/2227779X.

Carbon Budget. (2016). *The emission pledges from the US, EU, China, and India leave no room for other countries to emit in a 2°C emission budget (66% chance).* Retrieved from http://www.globalcarbonproject.org/carbonbudget/

Carbon Tracker. (2016). Carbon Tracker Initiative. Retrieved from http://www.carbontracker.org/

Climate Interactive. (2016). *Climate scoreboard update: National plans lead to 3.3°C.* Retrieved from https://www.climateinteractive.org/analysis/climate-scoreboard-update-national-plans-lead-to-3-4c/

Dietz, R., & O'Neill, D. W. (2013). *Enough is enough: Building a sustainable economy in a world of finite resources.* San Francisco: Routledge.

du Pont, Y. R., Jeffery, M. L., Gutschow, J., Christoff, P., & Meinshausen, M. (2016). National contributions for decarbonizing the world economy in line with the G7 agreement. *Environmental Research Letters, 11*(5), 054005. https://doi.org/10.1088/1748-9326/11/5/054005.

Hansen, J., Sato, M., Kharecha, P., von Schuckmann, K., Beerling, D. J., Cao, J., et al. (2017). Young people's burden: Requirement of negative CO_2 emissions. *Earth System Dynamics, 8*(3), 577–616. https://doi.org/10.5194/esd-8-577-2017.

IPCC. (2014). In K. Pachauri & L. A. Meyer (Eds.), *Climate change 2014: Synthesis Report. Contribution of Working Groups I, II and III to the Fifth Assessment Report of then Intergovernmental Panel on Climate Change.* Geneva: IPCC.

Kinley, R. (2017). Climate change after Paris: from turning point to transformation. *Climate Policy, 17*, (1), 9–15. https://doi.org/10.1080/14693062.2016.1191009.

Le Quere, C., Andrew, R. M., Canadell, J. G., Sitch, S., Korsbakken, J. I., Peters, G. P., et al. (2016). Global carbon budget 2016. *Earth System Science Data, 8*(2), 605–649. https://doi.org/10.5194/essd-8-605-2016.

Mackey, B., & Rogers, N. (2015). Climate justice and the distribution of rights to emit carbon. In P. Keyzer, V. Popovski, & C. Sampford (Eds.), *Access to international justice* (pp. 225–240). Abingdon: Routledge.

Mattoo, A., & Subramanian, A. (2012). Equity in climate change: An analytical review. *World Development, 40*(6), 1083–1097. https://doi.org/10.1016/j.worlddev.2011.11.007.

Meyer, A. (2000). Contraction and convergence. The global solution to climate change. Foxhole, Devon, UK: Green Books for the Schumaker Sociey.

Peters, G. P., Andrew, R. M., Canadell, J. G., Fuss, S., Jackson, R. B., Korsbakken, J. I., et al. (2017). Key indicators to track current progress and future ambition of the Paris Agreement. *Nature Climate Change, 7*(2), 118–122. https://doi.org/10.1038/nclimate3202.

Peters, G. P., Andrew, R. M., Solomon, S., & Friedlingstein, P. (2015). Measuring a fair and ambitious climate agreement using cumulative emissions. *Environmental Research Letters, 10*(10), 105004. https://doi.org/10.1088/1748-9326/10/10/105004.

Prinn, R. G., & Reilly, J. M. (2017). *Transparency in the Paris Agreement.* Retrieved from https://globalchange.mit.edu/sites/default/files/MITJPSPGC_Rpt308.pdf.

Robiou Du Pont, Y., Jeffery, M. L., Guetschow, J., Rogelj, J., Christoff, P., & Meinshausen, M. (2017). Equitable mitigation to achieve the Paris Agreement goals. *Nature Climate Change, 7*, 38–43.

Rogelj, J., den Elzen, M., Hohne, N., Fransen, T., Fekete, H., Winkler, H., et al. (2016). Paris Agreement climate proposals need a boost to keep warming well below 2 °C. *Nature, 534*(7609), 631–639. https://doi.org/10.1038/nature18307PMID:27357792.

Rogelj, J., Schaeffer, M., Friedlingstein, P., Gillett, N. P., Van Vuuren, D. P., Riahi, K., et al. (2016). Differences between carbon budget estimates unravelled. *Nature Climate Change, 6*(3), 245–252. https://doi.org/10.1038/nclimate2868.

Rose, S. K., Richels, R., Blanford, G., & Rutherford, T. (2017). The Paris Agreement and next steps in limiting global warming. *Climatic Change, 142*(1–2), 255–270. https://doi.org/10.1007/s10584-017-1935-y.

Schandl, H., Hatfield-Dodds, S., Wiedmann, T., Geschke, A., Cai, Y., West, J., et al. (2016). Decoupling global environmental pressure and economic growth: Scenarios for energy use, materials use and carbon emissions. *Journal of Cleaner Production, 132*, 45–56. https://doi.org/10.1016/j.jclepro.2015.06.100.

Skeie, R. B., Fuglestvedt, J., Berntsen, T., Peters, G. P., Andrew, R., Allen, M., & Kallbekken, S. (2017). Perspective has a strong effect on the calculation of historical contributions to global warming. *Environmental Research Letters, 12*(2), 024022. https://doi.org/10.1088/1748-9326/aa5b0a.

Stockholm Institute. (2016). *Putting the 'enhanced transparency framework' into action: Priorities for a key pillar of the Paris Agreement*. Retrieved from https://www.sei-international.org/mediamanager/documents/Publications/Climate/SEI-PB-2016-Transparency-under-Paris-Agreement.pdf

UNFCCC. (1992). *United Nations framework convention on climate change—Rio Declaration, Rio Janeiro*. United Nations.

UNFCCC. (1997). *United Nations framework convention on climate change: Kyoto Protocol*. Kyoto, Japan: United Nations.

UNFCCC. (2015). *Paris Agreement*. Paris, France: United Nations.

UNFCCC. (2016) Aggregate effect of the intended nationally determined contributions: an update. Retrieved from https://unfccc.int/resource/docs/2016/cop22/eng/02.pdf

Wang, T., & Watson, J. (2012). Scenario analysis of China's emissions pathways in the 21st century for low carbon transition. *Energy Policy, 38*(August 2009), 3537–3546. https://doi.org/10.1016/j.enpol.2010.02.031.

Winkler, H., Mantlana, B., & Letete, T. (2017). Transparency of action and support in the Paris Agreement. *Climate Policy, 17*, 1–15. https://doi.org/10.1080/14693062.2017.1302918.

Index

A
Adhi Borga, 181
Asia-Pacific region, 17
Australian enterprises, 52
Australian pledge, 28

B
Bangladesh Agricultural Development Corporation (BADC), 175
Bangladesh microcredit institutions, 181
Bangladesh Water Development Board, 173
Barind Multi-purpose Development Authority (BMDA), 176
Barind tract, 169
Biomass, 162, 163
Brundtland Report, 151
Business-as-usual scenario, 115

C
Cancún Agreements of COP16, 192
Carbon dioxide equivalent (CO_2e) emissions budget
　anthropogenic emissions, 211
　carbon pollution, 210
　climate change, 210
　climate mitigation, 209
　critical-safety approach, 210
　fair share, 214–216
　global economy, 209
　global safety net/global budget approach, 211
　greenhouse emissions, 212
　national commitments and resultant emissions trajectories, 212
　NDCs and INDCs, 212
　Paris Agreement, 210, 211
　political and economic growth, 211
　transparency, 216, 217
　Trump administration, 210
Carbon disclosure
　carbon emissions, 97
　CDP, 88, 93
　climate change, 91
　CO_2 emissions, 90
　emergence, 91
　FedEx and UPS, 89, 90, 96
　FedEx's approach, 95
　institutional statement, 95
　institutionalisation, 88, 97
　organisational practices, 91
　symbolic behaviour, 92
Carbon Disclosure Project (CDP), 88
Carbon space, 151
Centre for Study for Science and Environment, 161
China
　aggressive climate change strategy, 145
　economic growth, 130
　environmental issues, 144
　harmonious society, 130
　hi-tech development and service sector employment, 130
　legitimacy, 144, 145
　national economy, 130
　political legitimacy, 130
　quantitative and qualitative research, 145
　renewable energy, 130, 144
　scientific development, 130
　transformation, 130

China Dream, 130
Chinese Communist Party (CCP), 130
Civil society, 140, 141
Clean Energy Tax, 156
Clean Environment Tax, 156
Climate Action Tracker, 32
Climate Change Conference, 44
Coal, 158
Coal price trap, 143
Computable general equilibrium model, 137
Conference of the Parties (COP), 17, 151, 193
Corporatocracy, 44, 45
Critical-safety approach, 210

D
Department of Planning, 79
Dhaka Division, 177

E
Ecological modernisation (EM)
   climate change, 20
   climate change policies, 16
   concept, 16
   COPs, 17
   democratic decision-making, 19
   international negotiations, 17
   international policymaking, 17
   Kyoto Protocol, 17
   national and state government policymaking, 16
   social and economic issues, 16
   UNFCCC, 17
   versions, 19
Ecologically Sustainable Development, 21
Economic development, 142
Economic-demographic model, 105
Emissions-intensive investments, 28
Environment Protection Authority (EPA), 77
Environmental degradation, 134
Environmental impact assessment (EIA), 71
Environmental impact statement (EIS), 72
Environmental Planning and Assessment Act 1979 (NSW), 7, 64

F
Family planning programmes, 107
Fertility decline, 112
Fifth Assessment Report (AR5), 105
Fiscal incentives and disincentives
   informal subnational charges, 204
   national charges and taxes, 199–203
   national legislation and sovereignty, 192
   NTSR, 198
   REDD+, 192, 193
Fiscal instruments, 197
Forest Licence Fee, 198
Forest Resource Provision, 197, 198
Forestry fiscal instrument, 198
Fossil fuel and industry (FFI), 104
14th Finance Commission, 157
Fuel Sense programme, 96

G
Geneva Conventions, 31
Global budget approach, 218
Global neoliberal capitalism, 53
Global renewable energy investment 2004–2015, 35
Global warming, 150
Globalisation, 44
Governmentality, 46
Green Climate Fund (GCF), 153
Green revolution, 176
Greener economy, 131, 145
Greenhouse gas (GHG), 154
Greenhouse gas (GHG) emissions, 105, 114
   BASIX, 76
   climate mitigation, 67
   EIA, 71, 72, 77
   frameworks, 70
   legislative amendments, 79
   "net-zero" emissions, 63
   NSW, 63
   planning framework, 69
   reduction, 65
   SEPPs, 69, 74
   strategic planning, 68
Greenhouse Gas Protocol, 89
Groundwater depletion, 175

H
High-fertility countries, 114, 120
Human mobility, 170
Hydroelectricity, 163

I
Impressive French diplomacy, 134
Inconsistency, 134
India
   climate change and development
     agriculture and fisheries, 155
     biocapacity, 153

Index

CO_2 emissions, 154
  energy, 154
  greenhouse gas emissions, 154
  industry, 155
  LULUCF, 154, 155
  macro indicators, 153
  petroleum refining and solid fuel manufacturing, 155
  residential and commercial, 155
  transport, 155
  urbanisation, 153
emission intensity, GDP, 165
India's Mitigation Effort
  biomass, 162, 163
  energy consumption, 159–160
  energy sector, 157, 158
  greenhouse gas emissions, 157
  hydropower, 163, 164
  renewable energy, 161, 163
  solar energy, 161, 162
  thermal power, 158, 161
  wind energy, 162
Indonesian National Development Planning Agency, 195
Infinite economic growth, 52
Institutional Logics, 91
Intended Nationally Determined Contribution (INDC), 164, 165
  budgetary sources, 156
  carbon sink, 157
  climate mitigation and adaptation, 156
  domestic and international resources, 156
  fossil fuels (petrol and diesel), 157
  impediments
    demonetisation, 164
    oil price, 165
    US policy, 165
  perform achieve and trade, 157
  project identification and sanctions, 156
  renewable energy certificates, 157
  UNFCCC, 155
Intended Nationally Determined Contributions (INDCs), 2
Inter-Governmental Negotiating Committee, 151
Inter-Governmental Panel on Climate Change (IPCC), 17, 105, 150
International Energy Agency, 157
International political developments, 144
Investment signal feedback, 34
Investor-state dispute settlement (ISDS) mechanisms, 33

K
Kyoto Protocol, 17, 32, 35, 96, 216

L
Land Use, Land Use Change and Forestry (LULUCF), 154, 155
Liberalisation, 47
Liberalism, 46
Local Environmental Plans (LEPs), 69
Low-fertility projection, 106

M
Madhupur clay, 176
Materialist/postmaterialist value change, 140
Millennium Development Goals (MDGs), 107
Million Tonnes of Oil Equivalent (MTOE), 158
Ministry of Environment and Forest (MoEF), 153
Ministry of Water Resources, 173

N
National Adaptation Fund, 157
National Adaptation Plans for Action, 120
National Bank for Agriculture and Rural Development, 156
National Clean Environment Fund, 156
National Development and Reform Commission (NDRC), 138
National Electricity Policy, 161
National Energy Administration, 137–138
Nationally determined contributions (NDCs), 26, 194, 211
Natural resource constraints, 134
New South Wales (NSW), 6, 62
Nongovernment organisations (NGOs), 45, 173
Non-tax state revenue (NTSR), 197
Norway's Minister for Climate and Environment, 196

P
Paris Agreement (COP21)
  China's actions, 133
  climate-resilient development, 132
  developed and developing economies, 133
  domestic policy shifts, 132
  issues, 152
  low greenhouse gas emissions, 132
  nation's mitigation actions and targets, 152
  sustainable development, 152
Paris Climate Agreement of 2015, 3, 26

Paris Climate Summit, 32
People's Republic of China (PRC), 130
Policy priorities
    annual reports (2009 & 2016), 139, 140
    decision-making power, 137
    five-year plans, 138
    institutional conflict, 137
    policy contradictions, 137
    PRC's political and economic structure, 137
Policy shift
    capacity and growth, 136, 137
    domestic economic transformation, 134
    investment, 134, 135
Policy-based projections, 8
Political pressure feedback, 30
Pollution licence reviews, 79
Pollution licences, 78
Population Health and Environment programmes, 118
Population policy, 110
Population-energy-technology (PET), 104
Private sector experiences, 193
Public perception, 140, 141

R
Rajshahi District, 172
Ratchet mechanism, 27
Reducing emissions from deforestation (RED), 193
Reducing emissions from deforestation and forest degradation (REDD+), 9
Reforestation Fund, 198
REmax scenario, 137
Renewable energy investment, 34
Renewable Energy Law, 142, 143
Results-based payment approach, 194
RoSE project, 109
Rural–urban migration, 168–170

S
Seasonal drought
    agricultural irrigation, 175
    BADC, 175
    Barind tract, 169
    BMDA, 176
    critical drivers, 174
    cyclone or flood, 170
    decision-making approaches, 171
    environmental factors, 182
    factors, 169
    groundwater depletion, 175
    groundwater irrigation, 176
    individual characteristics, 171
    longitudinal studies, 169
    low-income countries, 171
    microfinance projects, 178
    migration, 168, 179, 180
    NGO staff, 178
    nonenvironmental drivers, 174
    nonmigrant families, 169
    outmigrants, 173, 174
    pragmatic shortcoming, 172
    rural–urban migration, 170
    social and environmental challenges, 168
    Tanore Sub-district, 172
    temporary migration, 170
    vulnerability, 174
Seasonal migrants, 178
Shared socioeconomic pathways (SSPs), 105
Soil degradation, 168
State and Federal Climate Mitigation Laws, 64
State environmental planning policies (SEPPs), 69, 81
Stocktake mechanism, 215–217
Strategic Release Framework, 70
Stress Threshold Model, 171
Sustainability logic, 88, 91
Sustainable development
    Asia-Pacific, 50
    climate change, 51
    corporatocracy, 44, 45, 48
    eco-efficient global economy, 45
    financial institutions, 47, 49
    financial markets and transportation, 44
    globalisation, 44
    governmentality, 46
    individualism, 51
    labour markets, 48
    lifestyle and behaviour changes, 45
    neoclassical approach, 52
    political financing, 49
    socioeconomic, 48
Sustainable Development Goals, 4
Sustainable Development Goals plan, 2, 152
Sustainable economy
    COP21 agreement, 3
    growth and prosperity, 4
    NSW, 7
    Paris Agreement, 2
    REDD+, 9
    transition, 4
    transportation industry, 7
    UNFCCC agenda, 2, 5
    zero-emission goal, 4
Sustainable Planning Act 2009, 67

Index

T
Tanore Sub-district, 173, 182
Total fertility rate (TFR), 111, 112
Transportation industry, 7
2009 United Nations Climate Change Conference in Copenhagen (COP15)
  climate protection, 132
  developing *vs.* developed world argument, 131
  differentiated responsibilities, 132
  economic development, 132
  poverty eradication, 132
  principles, 131
  sustainable development, 132
2015 Paris Agreement
  carbon-intensive infrastructure, 37
  collective progress, 31
  feedback mechanisms, 29
  global stocktake, 31
  greenhouse gas concentration, 26
  ISDS, 33
  lock-in effect, 37
  "market message" or "investment signal", 28
  political pressure, 28
  political pressure cycles, 30
  ratchet mechanism, 27
  "reinforcing" feedback, 29
2015 Paris Conference, 130

U
Ultra-super critical technology, 161
UN's population model, 107

United Nations Climatic Change Conference, 152
United Nations Framework Convention on Climate Change (UNFCCC), 17, 104, 150–152, 192
United Nations Sustainable Development Goals, 5
USAID study, 117
US-based carpet manufacturer Interface, 19

V
Victoria's *Climate Change Act 2010*, 66
Voluntary family planning programmes, 117
vulnerabilities assessment, 172

W
Washington Consensus, 47
Wind power
  concession bidding, 143
  demonstration phase, 143
  industrialisation phase, 143
  scale-up and localisation phase, 143
World Bank, 193
World Bank's Forest Carbon Partnership Facility, 192
The World Commission on Environment and Development, 17
World Resource Institute, 153
World Trade Organization, 31, 33

Printed in the United States
By Bookmasters